Dragonfly Behavior

Georg Rüppell • Dagmar Hilfert-Rüppell

Dragonfly Behavior

Discovering the Dynamic Life
of an Ancient Order of Insects

 Springer

Georg Rüppell
Zoologisches Institut
Technische Universität Braunschweig
Cremlingen, Germany

Dagmar Hilfert-Rüppell
Inst. für Fachdidaktik der Naturw.
Technische Universität Braunschweig
Braunschweig, Germany

The translation was done with the help of an artificial intelligence machine translation tool. A subsequent human revision was done primarily in terms of content.

ISBN 978-3-662-70233-8   ISBN 978-3-662-70234-5   (eBook)
https://doi.org/10.1007/978-3-662-70234-5

Translation from the German language edition: "Verhalten von Libellen" by Georg Rüppell and Dagmar Hilfert-Rüppell, © Autoren 2024. Published by Springer Berlin Heidelberg. All Rights Reserved.

Cover illustration: © Georg Rüppell

This Springer imprint is published by the registered company Springer-Verlag GmbH, DE, part of Springer Nature.
The registered company address is: Heidelberger Platz 3, 14197 Berlin, Germany

If disposing of this product, please recycle the paper.

To Jan, Maren and Olav

FOREWORD

There are more dragonfly species on earth than mammal species, yet many people only know "the dragonfly" as one species and when asked may remember that some were blue and others red. In contrast, we can all name a wide variety of mammals without giving it much thought: Dog, cat, mole, hedgehog, bat, deer, elephant, dolphin. Depending on their habitat, these familiar mammals differ significantly in the shape of their limbs: legs, fins, wings or other conspicuous features such as antlers, spines or proboscis. Dragonflies, at first glance, are quite uniform with four wings, six legs, a head with two large eyes. The wings differ somewhat, as do the body proportions and all with a dramatic range of stunning colours. The distinguishing features between different species and even families of dragonflies are often the realm of specialists, the odonatologists (dragonfly researchers).

So what is the difference between all the many dragonflies on earth and what accounts for their great biodiversity? One answer to this is perhaps the great variety of different behaviour displayed by these fascinating insects. Most of us do not even notice their many activities, as they mainly take place in flight and often at high speed that our eyes cannot follow.

Georg Rüppell and Dagmar Hilfert-Rüppell have been known for many years for making the hidden behaviour of dragonflies visible to a wide audience. They do this primarily by means of slow-motion filming, both in the scientific field and in popular science, with many appearances on public television. They have won numerous national and international awards for these films. Slow-motion film is particularly suitable for allowing us to engage directly in dragonfly behaviour. Georg and Dagmar also use still photography, to capture sequences and special behaviour, accompanied by excellent illustrations that allow us to understand flight manoeuvres and prey capture events, for example, and thus make them comprehensible.

This new book "Dragonfly Behaviour" represents an overview of many years of intensive observation of dragonflies illustrated with many stunning photographs. This gives us new insights into their flight, prey capture, courtship, mating and egg laying, and we learn about many behaviours and flight manoeuvres that have not yet been described in the specialist literature. This requires not only their mastery of camera technique, which I always admire, but also their infinite patience, which allows them to spend hours at the water's edge just to observe something new or to be able to photograph or film a behaviour once seen. "Dragonfly Behaviour" thus opens up a wide new view into the fantastic behavioural ecology of dragonflies and the many special abilities of this group of animals. The book also exudes the enthusiasm for the subject of dragonfly behaviour that I personally have always greatly appreciated – and which drove me to join Georg Rüppell's working group at the University of Braunschweig as a young student. I apologize for not being patient enough to observe dragonfly behaviour later on and instead preferring to fish dragonfly larvae out of the water.

Frank Suhling , Professor of Geoecology, Technical University Braunschweig

INTRODUCTION

Dragonflies have evolved from the time of the Cretaceous period over 300 million years ago to the present day. They accompanied the dinosaurs from their arrival 245 million years ago through to their demise 65 million years ago. The descendants of dinosaurs, today's birds, have continued to make life difficult for dragonflies. For them, dragonflies were and are nothing more than a source of food and energy.

Only now, after 300 million years, are there beings who understand the seemingly endless success of dragonflies and can admire these graceful and spectacular creatures – us humans. Back in ancient Egypt, dragonflies were symbols of speed, yet only in the last few decades have we discovered how dragonflies fly so quickly and perfectly. It was only when slow-motion cameras made it possible to observe their agility and rapid flight using the latest and sophisticated technology could the flapping of their wings and air flow around the bodies of flying dragonflies be visualized. Now it is finally possible to understand the moving life of these ancient insects. This is extremely important for understanding their role in ecosystems as well as providing us with a great source of pleasure and mental well being.

We too have fallen in love with dragonflies and would like to share the joy of dragonflies and insights into their lives with the readers of our book. There are many books on dragonflies. There are comprehensive works like the classic "Dragonflies – Behaviour and Ecology of Odonata" by Philip Corbet or the beautiful, detailed book "Die Libellen Europas" by Hansruedi Wildermuth and Andreas Martens. Although both books contain an incredible amount of details about the life of dragonflies as well as numerous photos, it is mainly through reading that one gets an idea of the behaviour of the animals. With this book, we want to show important behaviours of this fascinating group of animals in pictures. With the help of more than 300 photographs and explanatory drawings from the life of dragonflies, we want to introduce the reader to the wonderful and sometimes bizarre world of dragonflies, according to the motto "a picture is worth a thousand words". We cannot provide a complete overview of all dragonflies in the world, as there are about 6300 known species. Many have not even been discovered yet. Instead, we want to show interesting behaviours, such as flying, fighting, threatening, courtship, reproduction, catching prey, or being eaten, that we were able to discover through slow motion filming. Dragonflies are highly efficient hunters, preying on anything in the air smaller than themselves. Their aquatic larvae are in no way inferior: Underwater, they are voracious predators and prey on everything that moves in their environment. In this way, they are the first stages which develop in the larval skin to later become the spectacular flying adults that we see around our ponds, lakes and rivers. They compete with each other using breathtaking flight manoeuvres and fight with their legs, they catch insects with acrobatic flight manoeuvres, and crack solid chitinous shells with toothed jaws. We can watch females choose mates and escape male harassment by burrowing into vegetation or diving underwater to lay their eggs. Finally, the study of their mating behaviour has yielded important results for evolutionary research.

You will see a lot of that in this book. We didn't just want to produce movies and pictures, but also wanted to understand their behaviour, so we have spent many years researching it. Getting close to a small piece of nature is like visiting paradise in the midst of the hustle and bustle of our technical media world.

The photographs are the result of many joint excursions in the field with photography and film cameras. Unlike most other insects, dragonflies have a relatively limited range, often restricted to one pond where their life cycle takes place. This is where they can be observed and studied. As shown in this book, dragonflies can be used as model organisms for the study of adaptations in ecology and behaviour. Dragonflies are unique in many ways – with the great flying abilities of adults catching prey with their spiny legs, the effective catching of prey by the larvae using their labial mask, as well as the famous dragonfly wheel during mating – all have evolved exclusively in dragonflies.

There are many general biological principles behind the exciting behaviour of dragonflies, which we explain in the short texts accompanying the pictures. Of course, there is still a lot more research to be done, but that should not stop us from reporting on our findings now, so that everyone, of all ages, can better understand these no longer mysterious insects. We would also like to encourage our readers to go out and observe, photograph or video dragonflies for themselves.

In order to understand the role of our protagonists, the dragonflies, the reader must tolerate the fact that although most of the interpretations are based on scientific findings (see bibliography), some are our own interpretations. For the sake of clarity and simplicity of presentation, we have sometimes had to use the somewhat humanizing style that is sometimes common in behavioural ecology.

Georg Rüppell & Dagmar Hilfert-Rüppell, Braunschweig 2024

Some dragonfly behaviour can also be viewed directly in videos. In the text, the locations are marked with a symbol ▶ and a blue short link. By entering the short link in one of the usual internet bowsers, you will be redirected to the film. With the e-book just click on the link.

CONTENT

Naming

Dragonflies are separated into two groups: the smaller Damselflies and the larger and powerful true Dragonflies, in this book called Dragonflies. There are two vernacular names for the species in English, the older ones, also used by the British Dragonfly Society mainly in Britain and Ireland, and those proposed by Dijkstra and used in Europe in general. We take the names used in Europe because this book is primarily about European species. The corresponding scientific name makes confusion unlikely. The appendix contains a list of both English versions and the scientific names.

BODY

▲

This Blue Hawker (*Aeshna cyanea*, above) is one of the most common Dragonflies in Central Europe. It can also be found in almost every garden pond and grows to around seven centimetres in length. At first glance, no other insect looks as perfect a flier as a dragonfly: huge eyes in front of a bulging front body, the thorax. This chest section is the power house: the powerful flight muscles work in it. In large dragonflies, these muscles can account for up to 60 % of the body weight and powerfully propel the huge wings. The front and rear wings of Dragonflies are differently shaped, hence the scientific name Anisoptera (unequal-winged dragonflies).

The narrower front wings are more flexible than the wider hind wings, which are particularly responsible for more lift but also for effective braking. The tubular abdomen is a storage space for digestion, the reproductive organs and weight-reducing air sacs. It is also used for stabilization, like the long balancing rod of a tightrope walker.

© The Author(s), under exclusive license to Springer-Verlag GmbH, DE,
part of Springer Nature 2024
G. Rüppell, D. Hilfert-Rüppell, *Dragonfly Behavior*,
https://doi.org/10.1007/978-3-662-70234-5_1

All of this is contained in 600–1000 mg (i.e. up to 1 g) of body mass in the large dragonflies. In contrast, the Demoiselles (*Calopteryx* species) weigh only 100–150 mg and the small damselflies such as the Bluets (*Coenagrion* species) only 20–30 mg.

And even in these tiny creatures, incredibly complex behaviour takes place in a matter of seconds. How do large and small dragonflies behave? What happens when they meet? Are they adapted to different habitats and prey?

The relatively large mass of large dragonflies is arranged to give very little air resistance, as shown here by a Blue Hawker. The shape of the head and eyes present hardly any edges or protrusions to the airflow, and the middle and hind legs, are closely tucked up under the body. The front legs lean against the head from behind – they have special tasks.

▼

Most damselflies are much smaller and more fragile than dragonflies. This newly emerged Azure Bluet (*Coenagrion puella*), for example, is only about 3.5 cm long and weighs about 40 times less than a Blue Hawker. They have far fewer flight muscles than large dragonflies. Their four wings are uniform and very agile because they are stalked – good for manoeuvrable flight.

▶ sn.pub/rmo6kb

▲ This is what a diverse dragonfly habitat looks like: a river with small still waters such as ponds or oxbow lakes. They are home to both flowing water species and those that prefer still waters.

▲ Damselflies feel at home here in a standing water rich in aquatic and emergent plants. Some dragonflies also come here to lay their eggs.

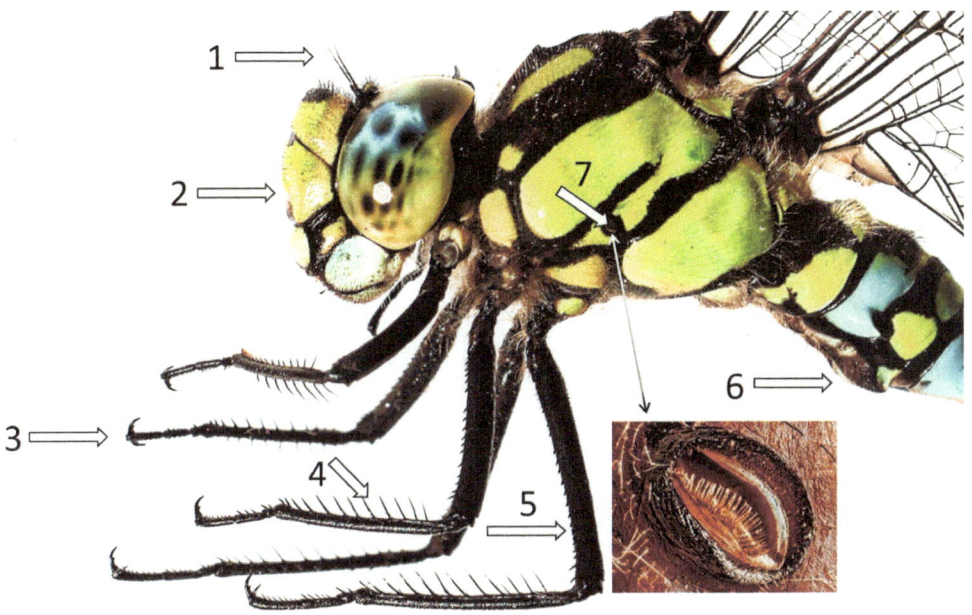

▲

Portrait of a Blue Hawker. The short antennae (1) measure airflow. The chitinous plates on the front of the head (2) act as low-resistance but shock-absorbing spoilers. The pointed claws (3) are the grappling hooks for landing and fighting. Long leg bristles (4) catch small prey, short bristles (5) hold large prey. The secondary copulatory apparatus (6) of this male is folded up. In the middle of the anterior body (thorax) is an opening, 'the spiracle' (7 and magnified below) of the respiratory tract (tracheal system).

EYES

Dragonflies look through several tens of thousands of tiny tubular individual eye cells. The Blue Hawker (*Aeshna cyanea*) has around 30,000, visible here as tiny dots that work together as compound eyes. The eyes meet at the front and top of the head. This gives them the best all-round vision and a high temporal resolution (about 7 to 10 times better than in humans). Dragonflies therefore see as we might do with our camera in slow motion.

▼

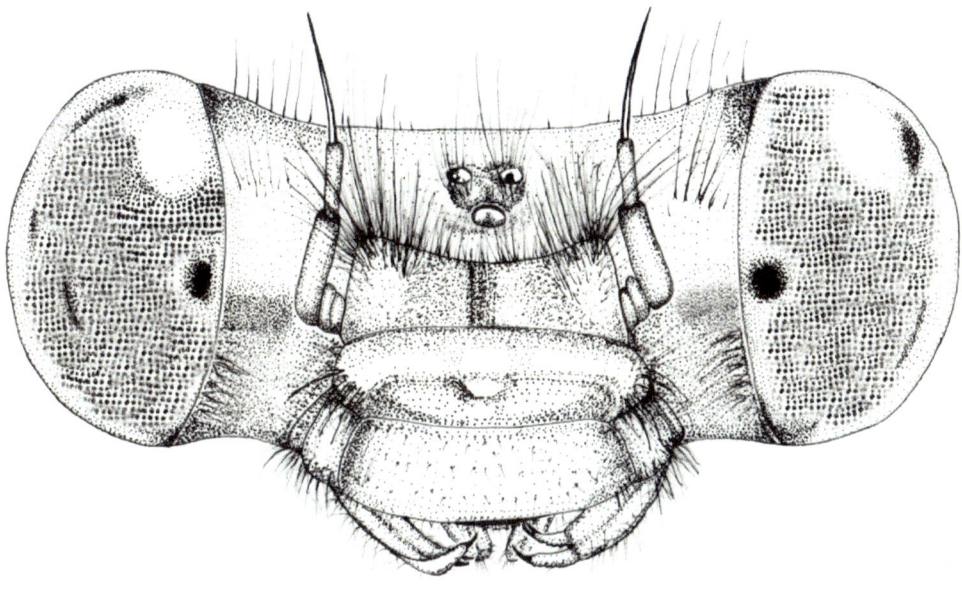

▲

Dragonflies are very visual animals. In addition to the huge, laterally positioned compound eyes, like all others this female Demoiselle (*Calopteryx*) has three small eyes (ocelli) in the middle of the head. They are important for light-dark vision and for flight position orientation with the aid of horizon detection. Dragonflies also have many other sensory organs such as the two antennae and the many hairs and bristles as mechanical and chemical sensors. Moreover, the hairs play a major role in keeping the body warm in many species and give rise to names like Downy Emerald and Hairy Dragonfly.

As with the Demoiselles and other damselflies, the compound eyes of Pincer-tails (*Onychogomphus*) do not meet at the top of the head. Together with the three ocelli and the antennae, they ensure a controlled flight. In almost all ma-noeuvres the head always remains horizontal. This is the only way the dragon-flies can safely direct their flight trajectories in the air.

▼

WINGS

▲

Dragonfly wings are a marvel of ultra-lightweight construction: large, light and yet rigid. They are arranged like corrugated sheet metal in a kink-fold structure. This increases rigidity and improves contact with the air flowing around them. Body fluid pulsates in the fine wing veins, which contain thin air channels. All this ensures stability and longevity – enabling large dragonflies to perform up to 130 million wing beats during an assumed 20 days of flight. Tropical species can do many times that. The uneven wing structure is clearly visible in the White-faces (*Leucorrhinia*) (above) and the male Beautiful Demoi-

© The Author(s), under exclusive license to Springer-Verlag GmbH, DE, part of Springer Nature 2024
G. Rüppell, D. Hilfert-Rüppell, *Dragonfly Behavior*, https://doi.org/10.1007/978-3-662-70234-5_3

selle (*Calopteryx virgo*) (below). In the latter, the many longitudinal veins and the very small and numerous wing cells are striking. Do Demoiselles have particularly demanding flight tasks?

▲

Dragonfly wings at different scales

Macro: Dragonfly wings are living networks. They consist of two ultra-thin membranes that overlap and between which there are stable longitudinal veins and fine transverse veins. Haemolymph (the blood-like body fluid) flows through these longitudinal veins and is drained off again via transverse veins at the rear edge of the wing.

During flight, small air vortices rotate in the "valleys" of the folds, helping to keep the air flowing around the wing. The wing veins are particularly powerful close to the body, as this is where the great force of the flight muscles is transmitted. Special stiffeners made of a high-molecular protein substance, resilin, are also found here. This rubber-like substance is incorporated at many intersections of the wing veins, is deformed under load and then snaps back into place so that the wing shape is maintained. A stiffening of the wing is shown at the top right.

(Stacked photograph of a dead Blue Hawker (*Aeshna cyanea*) was put together by stacking many individual images).

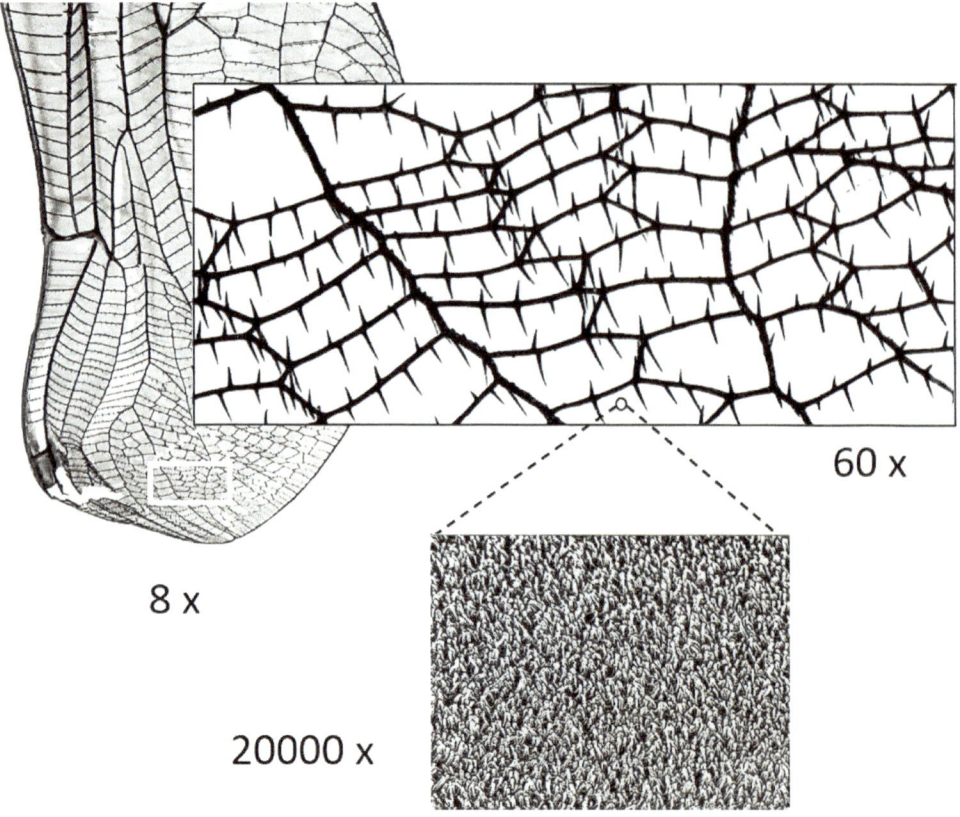

60 x

8 x

20000 x

▲

Micro (top): Magnified about 60 times, a section of the wing surface of the Blue Hawker looks like a stubble field. Spines protrude upwards and downwards on the wing veins. They help to create turbulence between the boundary layer of the wing surface and the air and thus also ensure an adherent airflow. One of our hypotheses suggests that they could also play a role in fights, similar to a knight's chain mail, because they are particularly long and numerous in the aggressive Blue Hawkers. The spines are also thought to act as spacers between the still moist wing surfaces when the dragonfly hatches. The distribution of the spines is interesting. Their number increases towards the wing tips and the rear edge. In the Blue Hawker, there are sometimes three times as many spines per unit area than near the body in the anterior wing area. This would suggest a function of the spines for airflow and defence. (Stacked photograph).

Nano (bottom): And this is what a surface section of a dragonfly wing looks like at 20,000x magnification taken with a Scanning Electron Microscope. An Australian-Japanese team of scientists has discovered this surface texture (photo Elena Ivanova) and found that it kills 70% of the bacteria that land on it. What clever hygiene precautions!

▲

Thanks to the special micro- and nano-structures, dragonfly wings are virtually unwettable. Dragonflies can still take off again even if they have fallen into the water or in the rain.

13

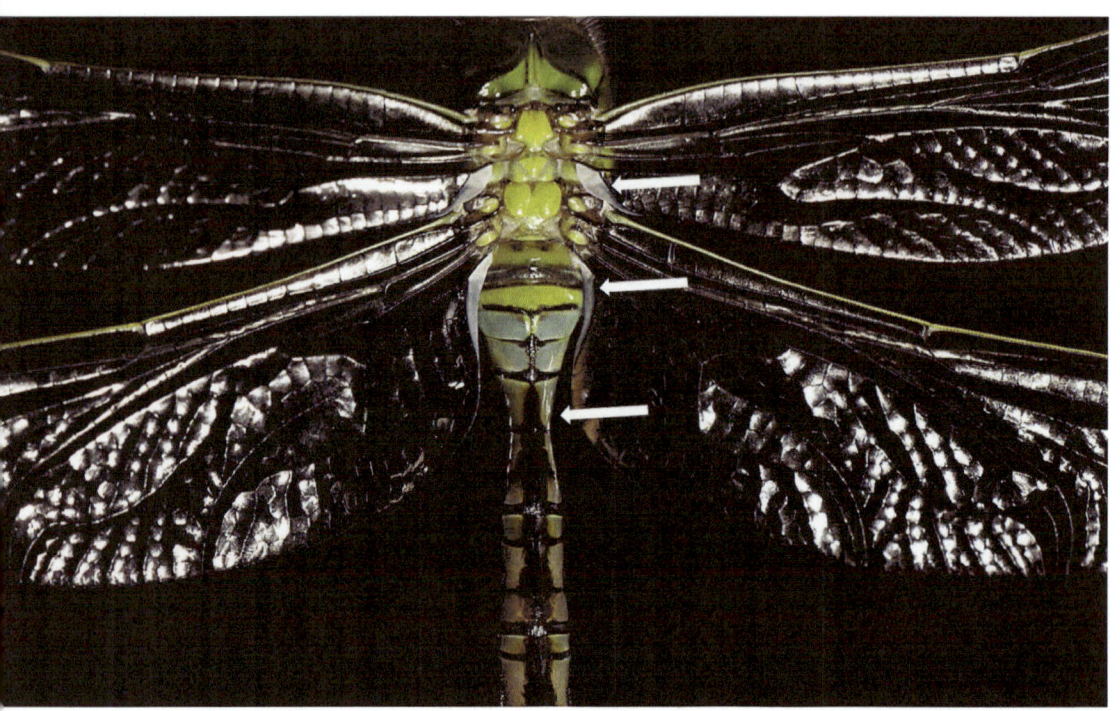

▲

In dragonflies, the wing layout is optimised for the size of the wings. The abdomen is constricted (lower arrow) so that the hind wings can be very wide without colliding with the body. However, in order to survive any contact without damage, very elastic yet firm rubbing strips are attached to both the hind and front wings (arrowed).

Differences in dragonfly wings

Forewings (top) and hindwings (bottom) of the Blue Hawker (*Aeshna cyanea*, left), the Azure Bluet (*Coenagrion puella*, top right) and the Beautiful Demoiselle (*Calopteryx virgo*, bottom right).

    The wings drawn by Richard Lewington in his and Dijkstra's identification book only show approximately the number and size of the wing cells. The photos, on the other hand, show the architecture of the wings, even if the light reflections are somewhat distracting. The Azure Bluet has the fewest wing cells at all four wings with around 600 on a total wing area of around 2 cm². The Blue Hawker has wings nine times as large with around 2000 cells all together and the Beautiful Demoiselle with half the wing area of the Blue Hawker has around 10,000 wing cells.

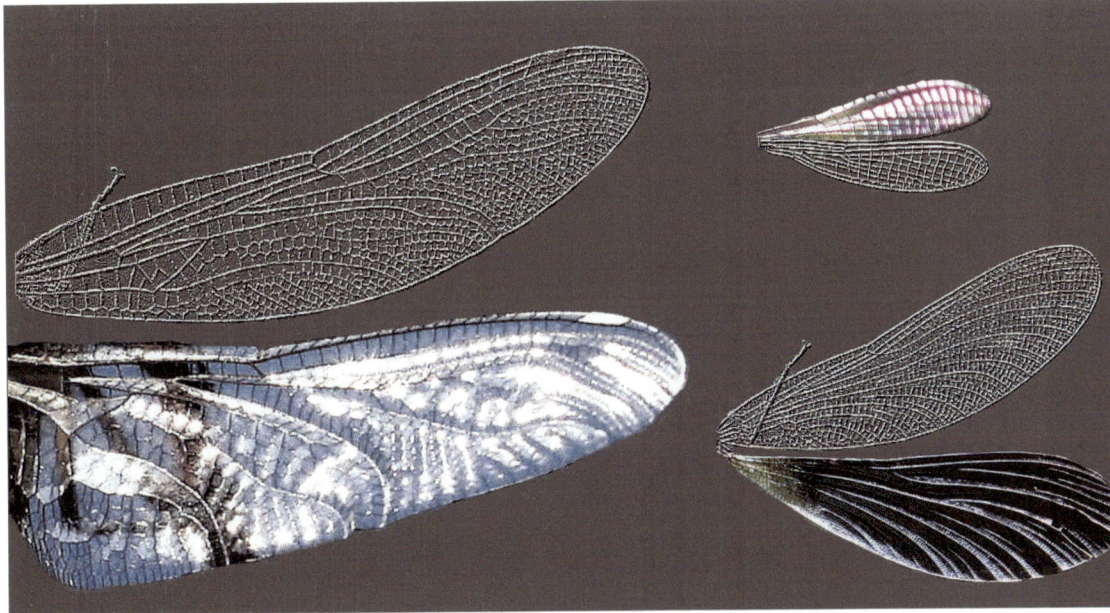

▲

Dragonfly wings, such as those of this Blue Hawker consist of many longitudinal and cross veins. Many strong longitudinal veins are located on the front sides of the wings. The body, shortly after hatching as shown here, has not yet reached its final colouration. The wing markings towards the tip of the wings are the pterostigmata (sing. pterostigma) (arrowed). Here they are still white, but later acquire a species-typical colouring. They are important as a counterweight for the wing beat, as they have an increased mass when filled with body fluid.

Insects have an open circulatory system with a tubular heart in the thorax that pulsates through muscles. The haemolymph flowing to the head only transports important substances for cell metabolism and immune defence, but no oxygen as in humans. Oxygen is transported directly to the sites of consumption via a respiratory channel system, the tracheal system. This only works well for insects when it is warm. Some insects like bumblebees have a special warm fur and will also warm themselves by vibrating. Dragonflies can do this to a certain extent, but they are not as well insulated as bumblebees. If it is too cold (or too hot) for too long, dragonflies cannot hunt prey and will therefore starve to death in these conditions.

THERMOREGULATION

Dragonflies are ectothermic animals. It is very important for them to quickly reach operating temperature in the morning so they can become active. These Common Bluetails (*Ischnura elegans*) snuggle closely to plant stalks, where it is almost eight degrees (Celsius) warmer than in the open air. (The arrows pointing outwards, above indicate the temperature between the stalks). It is also through such thermoregulation that this species is so successful in extending its activity to cool weather.

▶

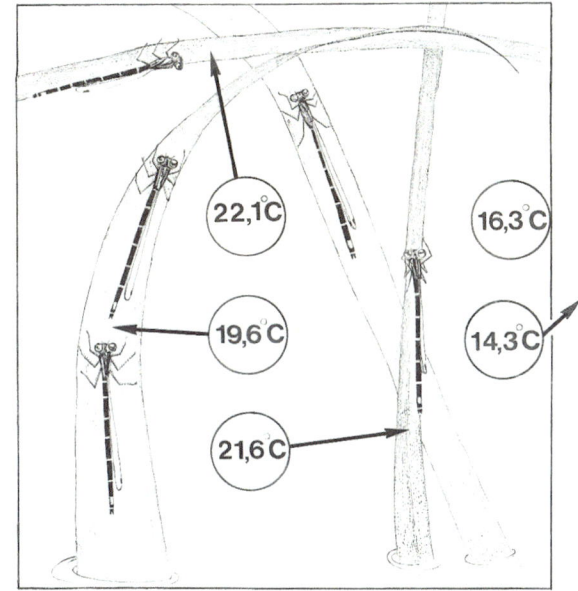

◀

To prevent overheating of their flight muscles, some dragonflies can dissipate the heat with the body fluid passing into the abdomen, from where it then radiates to the environment. Photo from below.

(*Aeshna cyanea*, male)

G. Rüppell, D. Hilfert-Rüppell, *Dragonfly Behavior*, https://doi.org/10.1007/978-3-662-70234-5_4

◄

Many dragonflies spend the night together (left: Western Demoiselles (*Calopteryx xanthostoma*)). This helps to protect each of them individually from predators. In the morning and evening, Demoiselles open their wings to capture heat and thus remain active for longer.

▲

On the left male and female Banded Demoiselles (*Calopteryx splendens*) and on the right a young male Copper Demoiselle (*Calopteryx haemorrhoidalis*) are warming up.

18

Banded Darters (*Sympetrum pedemon-tanum*, below) also roost together. However, like all true dragonflies, they always have their wings spread open throughout their lives.

The thorax and abdomen of this newly hatched Four-spotted Chaser (*Libellula quadrimaculata*) are almost transparent and allow heat radiation to penetrate their body. The air sacs in the body also warm up quickly and store heat. Perhaps this is one reason for their more northerly distribution. This species occurs in a wide belt of Eurasia and North America, crossing the Arctic Circle. Further south, it is found at cooler, higher altitudes.

19

AERIAL ACROBATS

Dragonflies like this Blue Hawker (*Aeshna cyanea*) are the best flyers among insects: with high agility they can fly sideways, backward, or even fly in an inverted body position. They can make rapid changes in flight directions with very high acceleration. In addition to hours of continuous flying – such flight characteristics seem unreachable for technical flyers▼

The explanation of the unique flight ability of dragonflies comes from flow research. With ever more sophisticated technology for the visualisation and calculation of air currents and flow, it has been found that bumblebees, for example, can indeed fly at temperatures which mathematically were denied by experts thirty years ago. Nobody back then could have known that small flying animals with movable wings generate vortices that they themselves can use for flying. This effect is greater, the smaller the wings are. Dragonflies fly in an area where often these vortices play the main role in the generation of air currents. If you look at the air flow and turbulence of a dragonfly in flight either on the spot or in a curve, you are confronted with pure chaos. Flow researchers can now calculate the detailed flight parameters and the vortex effects from these experiments. We now understand much more about how dragonflies can fly!

G. Rüppell, D. Hilfert-Rüppell, *Dragonfly Behavior*,
https://doi.org/10.1007/978-3-662-70234-5_5

Vortex patterns, which a dragonfly (here in red) generates in flight are made visible by the research of Haibo Dong at the University of Virginia. It is understandable that such complex flow images remained closed to science for a long time. This simplified view helps to interpret the flow chaos. In order for the dragonfly to stay in the air and not stall, a downward directed air jet must be generated, which can also be recognized here as the sum of the downward vortices. In addition, the leading and trailing edges of the wings generate vortices when flipping from one direction of movement to the other, which generate beneficial air forces directly on the wings.

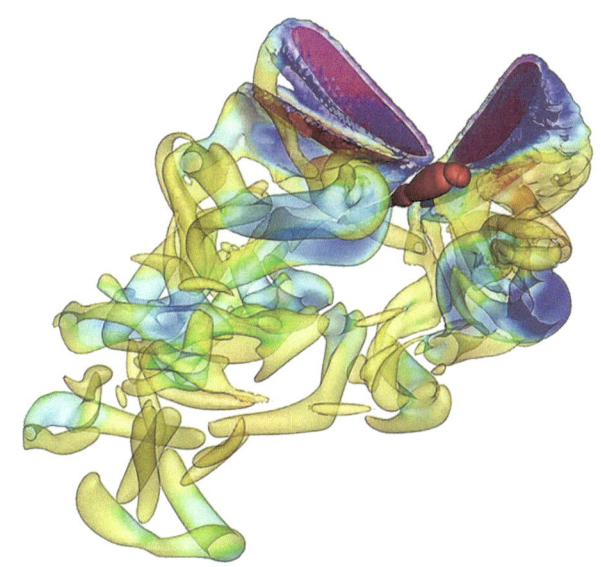

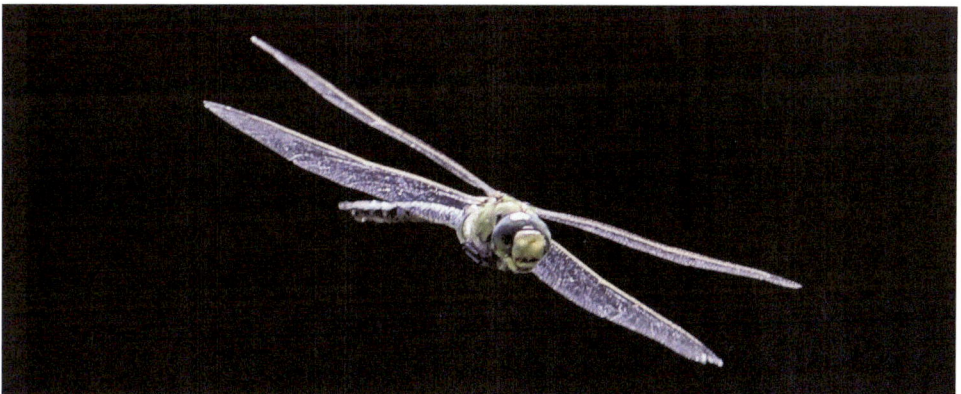

When gliding, the air movements are more pronounced. In large dragonflies, the front and rear wings are at different angles to the airflow coming from the front. This deflects the air downwards, which is important for lift.

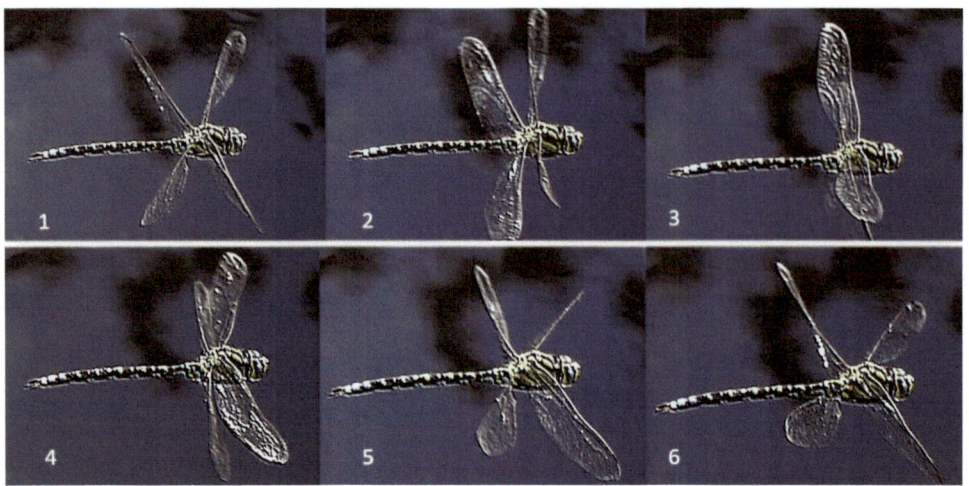

▲ How do dragonflies such as this Blue Hawker beat their wings when flying straight ahead? From the top back to the bottom front (e.g. like the hind wings in pictures 1 to 2) the wings then turn over and now strike upwards again with the underside on top (picture 4 to 6). Vortices are created at the reversal points, which are partially used to generate air forces. The hind wings lead this flapping movement and the front wings follow with a delay. It was found this flapping pattern to be the most energetically efficient for dragonflies because the air vortices of the forewings then have a positive influence on the air forces of the hindwings.

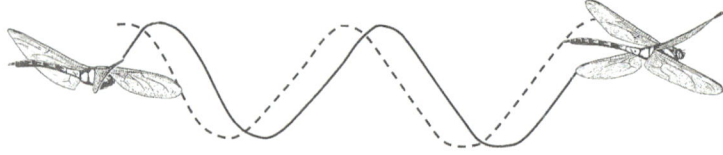

▲

The phase relationship of the front wings (dashed line) and the rear wings (solid line) during the straight flight of a hawker dragonfly can be represented graphically. Downward curve = downstroke, upward curve = upstroke.

If the hind wings are slightly ahead of the front wings, as shown here, they have a favourable influence on the flow around them, so that the dragonfly can fly for hours.

◀ The temporal delay of the forewings compared to the hind wings can be seen in this sequence of images of a hovering Blue Hawker (from top to bottom). In the upper left picture, the hind wings are already striking downwards, while the forewings are still striking upwards.

In the two following pictures below, however, the hind wings are already moving upwards and the forewings follow later.

▶ Dragonfly wings do not move rigidly like boards, but deform during the stroke. The wings twist in the process (like in the picture below) and therefore generate more and different air forces at the ends than at the base, also because they are exposed to a much faster flow there.

23

▲ The angles of attack of the wings can only be estimated in this Blue Hawker when viewed directly from the front or side. In the picture above, the hind wings are set very steeply against the stroke direction at the end of the stroke. They will immediately swing around so that the leading edges will again lead the movement forward.

▲ Top view of the thorax of a Blue Hawker: The wing joints of dragonflies not only look complicated, they are, as H. K. Pfau has shown. They consist of chitin braces, to which the flight and setting muscles attach. The main forces are transmitted to the wings by direct muscle action. However, there are also some indirect muscle effects when dragonflies flap their wings. These indirect flight muscles are tensed by deformations of the thorax.

▲ Dragonflies rarely use full wing strokes. This Blue Hawker only uses a medium stroke angle range, as the shot with a relatively long exposure time (1/200 sec) shows. However, she can double it if necessary.

The White Face (*Leucorrhinia*, below) is also just beginning its downstroke with its front wings not far up. ▼

25

▲

Damselflies, such as these newly hatched Azure Bluet (*Coenagrion puella*), flap their hind wings so far back that they often touch each other.

When the wings are subsequently flapped, air flows into the resulting gap, which is thought to generate vortices and favourable air forces and was described as a "clap and fling" mechanism by Weis-Fogh.

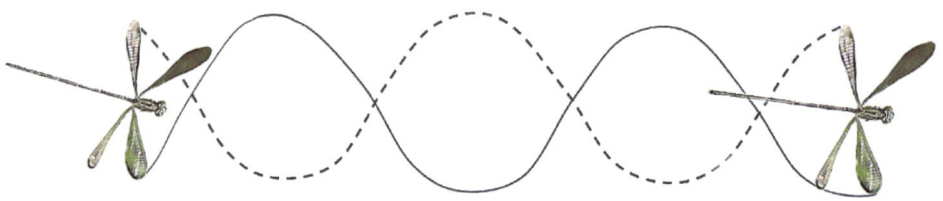

▲

Damselflies often beat their front wings (solid line) and rear wings (dashed line) in opposite directions when flying straight ahead and on the spot. This results in a uniform speed because one pair of wings alternates with the other in producing propulsion. However, during special manoeuvres such as flying backwards, they can also change this counter-rotating flapping.

▲

These two Azure Bluets move their wings differently: while the male (top) has been swinging its front and rear wings in opposite directions for several seconds, the female flaps both pairs of wings downwards at the same time right at the start, generating a particularly large amount of lift.

CURVING AND TURNING FLIGHT

In order to fly in a turn curve or when compensating for gusts of wind, dragonflies constantly twist their thoracic section back and forth with their wings – a finely responsive and precision system.

By twisting their bodies with their wings, these Blue Hawkers (*Aeshna cyanea*) fly in a right turn. The heads remain in a horizontal position. The air forces generated by the wings are now directed towards the centre of the turn curve to compensate for the inertial force acting outwards, and the steeper the body of the dragonfly when turning, the more this is done. This is particularly true of the dragonfly in the right-hand picture. Its body with the wings is extremely steep. The turn curve it flies is correspondingly tight and fast.

▼

▲ The bodies of these Blue Emperors (*Anax imperator*) are also inclined together with the wings during turning, while the heads are also held horizontally. Different inclinations and angles of attack of the wings result in differently narrow curves. Left: tight; center: wide; right: very tight.

▲ A male Black-tailed Skimmer (*Orthetrum cancellatum*) flies a steep turn to the right (in the direction of flight, phases from left to right). The inclined position and individual wing positions, such as that of the right hind wing, which is set at a steep angle during the downstroke in phase three, turn the dragonfly.

29

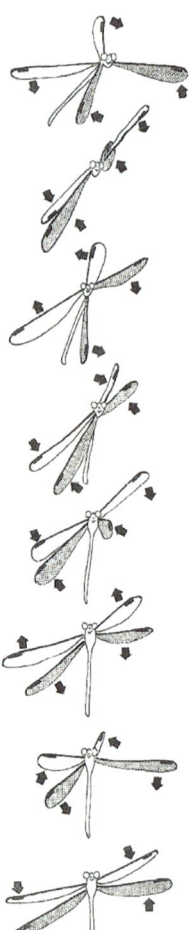

A good turn curve technique is part of the aerial combat of these Black-tailed Skimmers. The lower male is a bit late and always has to do all flight manoeuvres better than the top male to prevent him getting closer. (Phases from left to right).

◀ A Common Spreadwing (*Lestes sponsa*) flies a right curve: The rotation is caused by the inclination of the body and the wide swinging of the curve-outer (left) wings, recognizable by their strong shortening at impact. (Forewings are shown as transparent, hindwings hatched).

Turn of the Blue Hawker (*Aeshna cyanea*) while hovering on the spot
The dragonfly flies a 180° turn curve on the spot due to a slight inclination and different angles of attack on the two sides of the body. Steeply inclined wings in the outer curve

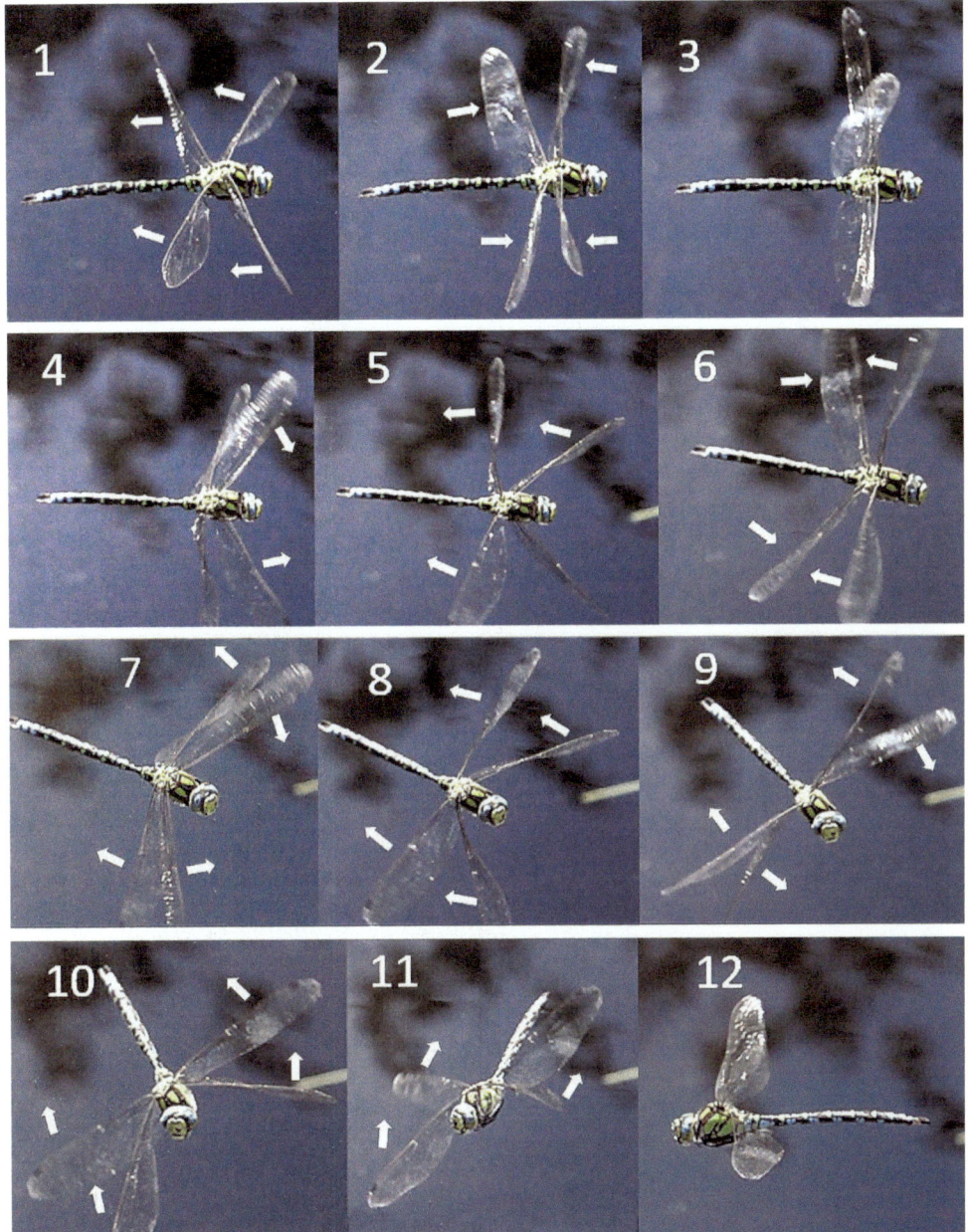

on the upstroke: 1, 5, 8; steeply inclined wings in the inner curve on the down-stroke: 2, 6, 9. The respective wings on the other side of the body are always less inclined. ▶ sn.pub/hkcov1

TAKE-OFF

   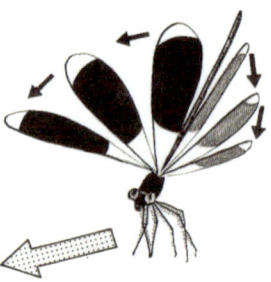

▲ Unlike grasshoppers, Demoiselles only push off slightly with their legs when taking off. They accelerate with their large wings to get into the air. On the right a Demoiselle is taking-off by a right turn.

◀

Skimmers (*Orthetrum*), weigh around three times more than Demoiselles and fly much faster. It is therefore not surprising that they have thicker legs, which they also use to push themselves off during take-off. The main acceleration comes from their wings (from bottom to top).

G. Rüppell, D. Hilfert-Rüppell, *Dragonfly Behavior*, https://doi.org/10.1007/978-3-662-70234-5_7

ACCELERATION

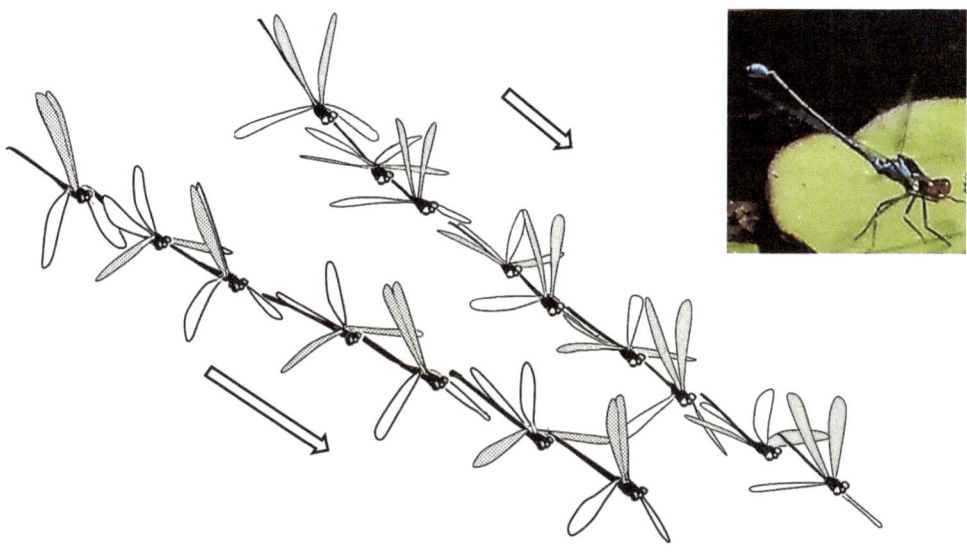

▲

The lower male of the Small Redeye (*Erythromma viridulum*) pursuing the upper one is about 20 centimetres behind. To catch up, it increases its stroke movement compared to the pursued by wider flapping (120° instead of 80° like the upper male) and a higher stroke frequency (62.5 Hz instead of 54.5 Hz) and thus achieves a speed of 3.7 m/s instead of 2.2 m/s like the pursued.

▶ sn.pub/atmoxb

BACKWARDS FLIGHT

This Banded Demoiselle (above) and the Small Redeye (below) both fly backwards with stroking both pairs of wings set at steep angles on a horizontal trajectory forwards (Demoiselle 2nd and 4th phase from left to right) and backwards at shallower angles (Demoiselle 3rd). ▼

The common redstart (*Phoenicurus phoenicurus*) flies backwards in a similar way. On the right, when flying straight ahead, the wings are flapped downwards, whereas when flying backwards (left) they are flapped horizontally forwards with their full broadside (= large angle of attack of the wings to the flapping wind now coming from the front). ▼

G. Rüppell, D. Hilfert-Rüppell, *Dragonfly Behavior*,
https://doi.org/10.1007/978-3-662-70234-5_9

INVERTED FLIGHT

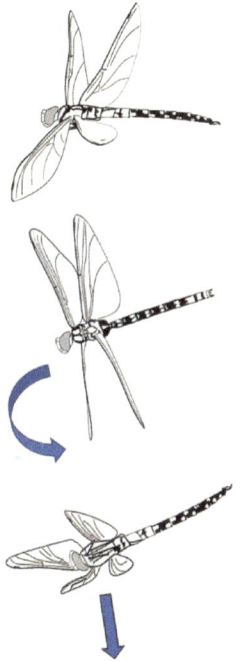

▲

A male Blue Hawker (*Aeshna cyanea*) escapes an attacking conspecific rival (white ovals in the photos on the right) by using an unusual avoidance tactic (image sequences from top to bottom): it turns on its back and thus accelerates strongly downwards. It reaches a downward speed of 3.4 m/s (bottom left image) in 0.07 seconds from flying on the spot. It achieved this by benefiting from the earth's gravitational pull and also by flapping its wings upwards to move air upwards. By setting the wings on the left side of the body so steeply that no

G. Rüppell, D. Hilfert-Rüppell, *Dragonfly Behavior*, https://doi.org/10.1007/978-3-662-70234-5_10

more lift was generated here, this male was able to turn quickly (drawing at left, middle picture).

 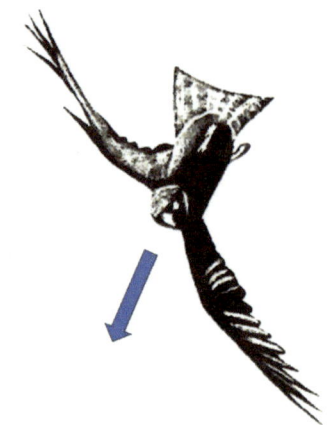

▲

A Black Kite (*Milvus migrans*) flies inverted just like the dragonfly when competing for prey and thus accelerates strongly downwards.

COLORATION

In most dragonflies, the males are strikingly coloured, while females often look camouflaged or cryptically coloured. The colours we see can be formed by pigments or as a result of light refraction on the surface (structural colours). Some dragonflies like this Downy Emerald (*Cordulia aenea*), have both pigments and structural colours, the ratio of which is not yet fully understood.

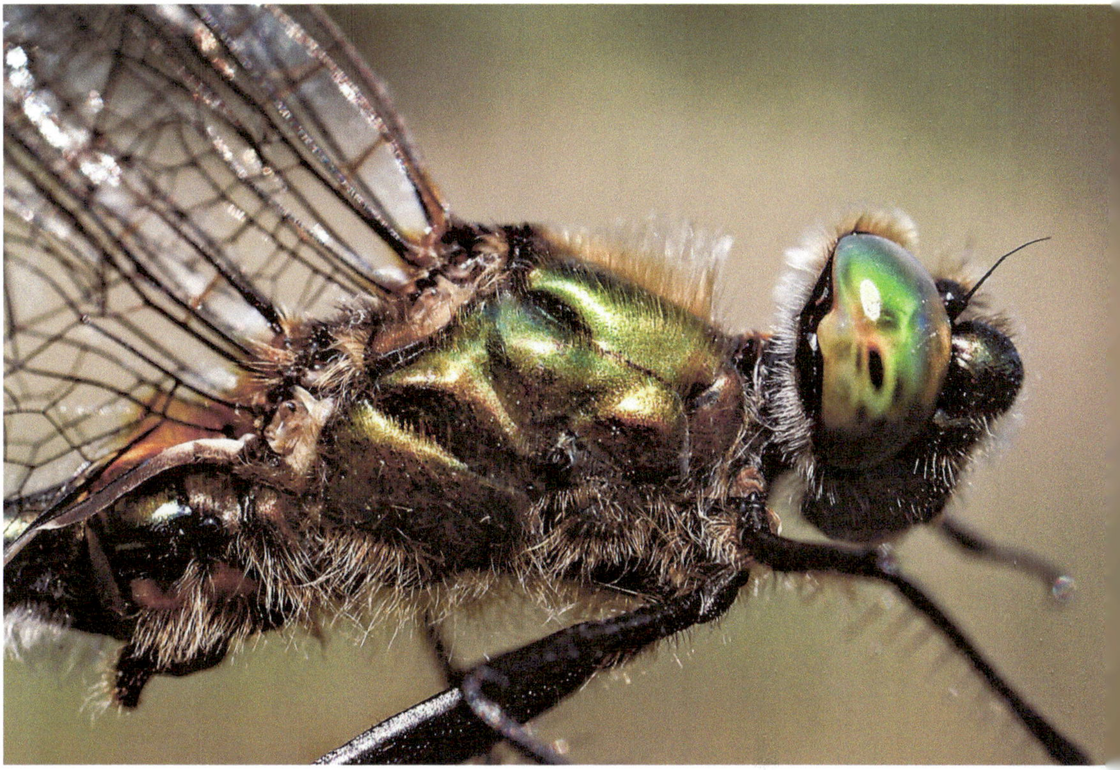

A thin layer of wax, the so-called pru-inescence, can also produce structural colours, here in light blue of this male Black-tailed Skimmer (*Orthetrum cancellatum*). Females hold on to the male's abdomen when mating, as shown here.

In doing so, they scrape off some of the pruinescence each time. Such darker scraped areas indicate many matings and an evidently successful male.

▶

◀

On the other hand, this male Darter (*Sympetrum*) glows red as a result of its pigment colours.

Not only are the bodies of many male dragonflies very brightly coloured – sometimes their wings are too, as shown in this male Banded Demoiselle (*Calopteryx splendens*). These wings serve not only for flying, but also for communication.

Structural colours can also be recognized by the fact that they are very different depending on the direction of the incident light. The blue of the wings in male Banded Demoiselles is such a structural colour. It ranges from light blue (left) to medium blue (centre) to blue-black (right) in this male.

The wing colours of this Giant Damselfly (*Megaloprepus caerulatus)*, from Central America (above) and those of the colourful Butterfly Dragonfly (*Rhyothemis fuliginosa),* from South-East Asia (left), are created by light refractions and can vary greatly depending on the angle of incidence of the light: from golden to deep blue-black.

A sitting male (1) of the South Asian metal-winged damselfly (*Neurobasis chinensis*) displays its wings while threatening its incoming rival (2, below). These structural colours shine very differently in the same animal, depending on how the light is reflected.

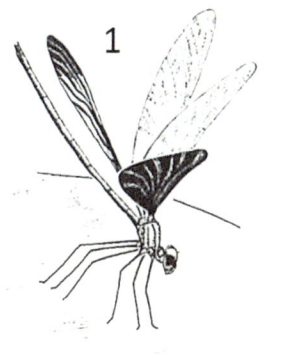
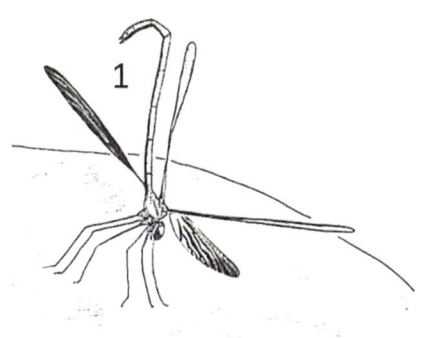
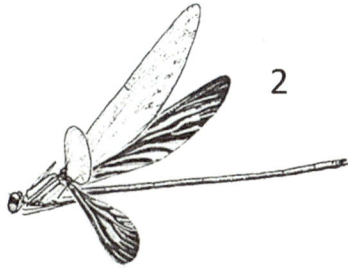

COLOURED WINGS BEAT DIFFERENTLY

Sexual selection has led to the development of coloured wings in dragonflies and damselflies. Females therefore have the opportunity to evaluate males before mating. Males with coloured wings not only use them for flight, they also send out signals, as in this Beautiful Demoiselle (*Calopteryx virgo*). This signalling effect has resulted in a completely different way of flapping their wings: in contrast to all dragonflies with clear wings. Demoiselles flap all four wings up and down at the same time during normal straight flight. However, they can also hold all four or two wings still for a moment, or even just one wing. Furthermore, during courtship flight they switch back to a completely different flight style, now high-frequency and out of phase. This diversity of flight has behavioural consequences.

In addition, Demoiselles generate a lot of air forces with the simultaneous beating because now four wings can move much more air than just always two, as would be seen in clear-winged dragonflies, whose fore and hind wings usually beat out of phase. Demoiselles can only maintain this flight style in exceptional cases. In this sequence of images, a male of the Beautiful Demoiselle flies a left curve.

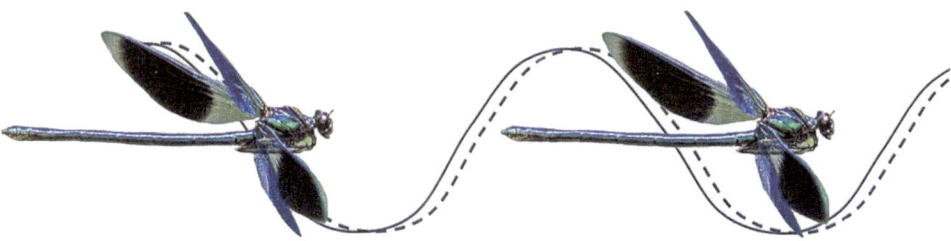

The simultaneous downward (downward curve) and upward (upward curve) flapping of the Demoiselles. The forewings (solid line) are a little earlier than the hindwings (dashed line).

43

▲ The upper reaches of streams and rivers are the habitat of Beautiful Demoiselles.

In relation to their body weight, Demoiselles have very large wings. This enables them to generate a lot of flight power and forward thrust. They can travel almost 16 cm with one wing beat, four times as far as other small damselflies such as Azure Damselflies. But they are also susceptible to wind gusts. This Banded Demoiselle (*Calopteryx splendens)* is knocked over by a strong gust of wind (above, from bottom right to top left). In the 4th phase, the male flips over, but is soon able to straighten up again. ▼

The females of the Demoiselles are inconspicuously camouflaged. They can also beat all four wings simultaneously together like the males. This means they can also fly quickly and often use this to escape the unwanted attention from the males.

Thus, Demoiselles fly backwards and forwards. The dashed arrows indicate the beating directions of the wings, the open arrows the flight directions of the damselflies. Same grayscale

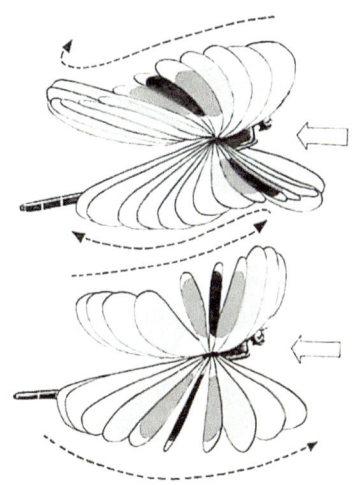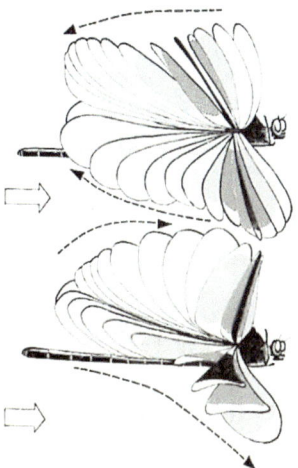

means simultaneous wing positions. When flying backwards (left), they set the wings steeply during the downstroke (bottom left), when the Demoiselles fly forwards (right) the wings are steeply set during the upstroke (top right).

In the rainforest of Panama, it is usually calm. Here, even larger wings can be moved, such as those of the largest damselflies (and dragonflies!) in the world, the Helicopter Giant Damselflies *Megaloprepus caerulatus* with a wingspan up to 19 cm. Flashes of light from the contrasting deep blue and white

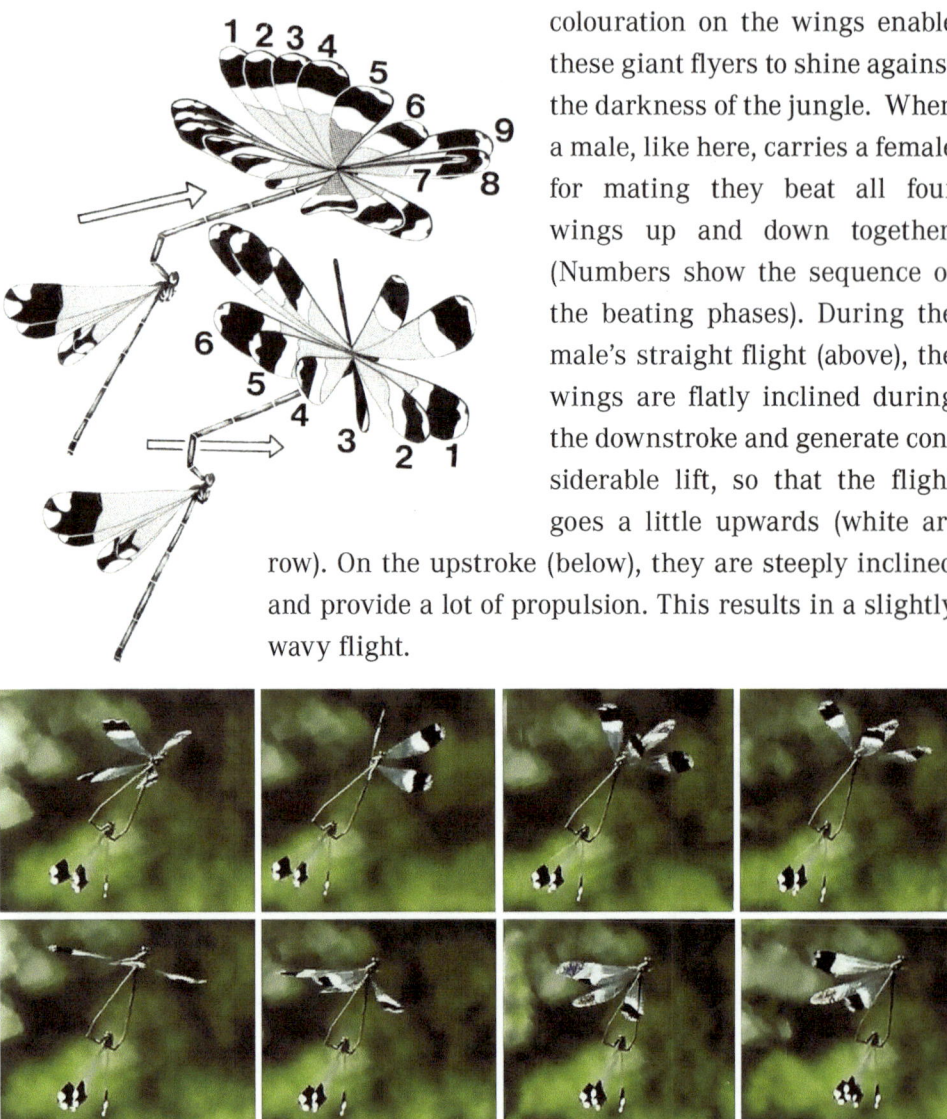

colouration on the wings enable these giant flyers to shine against the darkness of the jungle. When a male, like here, carries a female for mating they beat all four wings up and down together. (Numbers show the sequence of the beating phases). During the male's straight flight (above), the wings are flatly inclined during the downstroke and generate considerable lift, so that the flight goes a little upwards (white arrow). On the upstroke (below), they are steeply inclined and provide a lot of propulsion. This results in a slightly wavy flight.

▲ When this large, glowing pair of Helicopter Giant Damselflies flies through the jungle, one feels transported to an ancient time. During hovering flight (starting top left), it is easy to see the male's wings reversed at upstroke (bottom row) and stroking backwards on an almost horizontal path with the underside up. The smaller female does not use wing strokes and lets herself be carried along.

LANDING

The landing of the light Demoiselles is unspectacular. They simply fly with their legs outstretched just before landing, which then only bend slightly, rock forward a little and quickly come to rest.

In contrast, Black-tailed Skimmers (*Orthetrum cancellatum*) are relatively heavy and fly fast, they land with only four legs. The front legs are not used and are held behind the head – perhaps as an additional stabilizer for the heavy head?

G. Rüppell, D. Hilfert-Rüppell, *Dragonfly Behavior*, https://doi.org/10.1007/978-3-662-70234-5_13

DANGEROUSLY COLOURED WINGS

Coloured, dark wings can also have life threatening disadvantages: When it is very hot in the south of France, the dark wings of Demoiselles heat up so much that they are unable to hunt and need to perch in the shade and slowly starve to death (below).

During one August, only a few dozen of around 2000 Copper Demoiselles, (*Calopteryx haemorrhoidalis*) survived on the course of a small river of two kilometres. This is a trend that will continue to increase with increasing solar radiation due to global warming. To compensate the Demoiselles may in time shift their range.

In the even hotter southern regions, there are even species such as *Calopteryx syriaca* and especially *Calopteryx exul*, whose wings are no longer dark but hyaline (transparent) and therefore do not absorb as much heat.

▲

High population numbers of Copper Demoiselles in June at a small watercourse in Southern France (left) and a shady stream as an alternative habitat? (right).

G. Rüppell, D. Hilfert-Rüppell, *Dragonfly Behavior*, https://doi.org/10.1007/978-3-662-70234-5_14

In August, five females and three males of the Copper Demoiselle survived the midday heat in the shade. Many died of starvation and floated dead on the water.

▼

Prey Capture

Copper Demoiselles (*Calopteryx haemorrhoidalis*) in the south of France during their morning search for prey. ▼

G. Rüppell, D. Hilfert-Rüppell, *Dragonfly Behavior*, https://doi.org/10.1007/978-3-662-70234-5_15

Another special feature of dragonflies is their ability to catch prey with their legs which have many pointed spines and bristles. Demoiselles have around 600 of them and some are very long. Folded into a catching basket, they use their legs to catch even the smallest flying insects from the air. This requires the highest sensory ability and manoeuvring skills, as the prey can react at lightning speed. Dragonflies, on the other hand, are often successful thanks to large nerve cells and direct connections from the eyes to the flight muscles. Under controlled conditions in the laboratory, without the many disturbing factors, large dragonflies can achieve very high catch rates. In the field, where there are many distracting stimuli and other factors they tended to achieve lower catch rates.

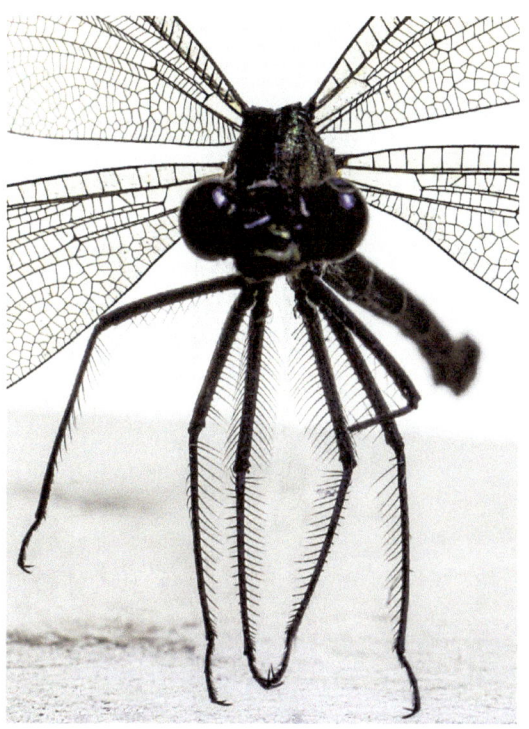 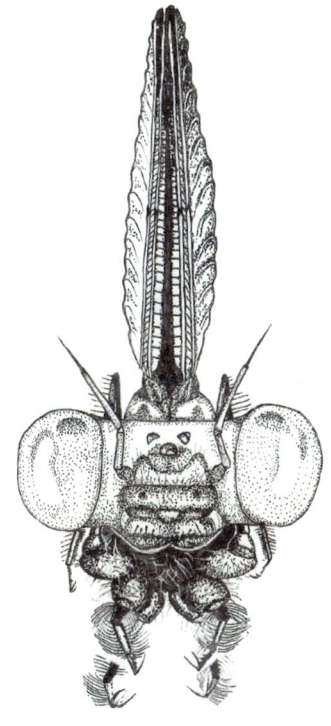

▲

From one moment to the next, the female Banded Demoiselle (*Calopteryx splendens)* unfolds its huge wings and spreads its bristle-covered legs: often an inescapable obstacle for its prey.

The catching basket of the dragonfly legs in action. Left, a Banded Demoiselle female has caught a small insect (two images composited into one), right a male makes a successful catch. Whilst still in the air, the prey is brought to the mandibles and eaten.

▼

▲

A male *Calopteryx splendens* shortly before catching prey. The projection of the spatial paths of the tips of the right hindwing (bottom) and the left forewing (top) to 2 (top) are shown (starting on the left). It takes 0.004 s from point to point. The rear wings are flapped close to the dragonfly's centre of gravity. They are the main motor and generate lift and propulsion rhythmically. The front wings move above the body and, in addition to producing lift and propulsion, are particularly important for controlling flight manoeuvres. They are sometimes even held still during the threatening flight of damselflies and then serve as signal flags. The dashed, diagonal lines mark the ends of individual wing beats.

▲

A Banded Demoiselle female (left) and a male (right) catch small insects with their legs (from top to bottom). Especially in the early morning, the mostly tiny prey animals contrast brightly, backlit against the background. Then you can often see the Demoiselles hunting against the light. Since most small insects fly very fast, the damselflies try to catch them when they come towards them.

▲ The race for prey of two Demoiselle females. Already at the start (1), the legs are spread, in the middle of the race (2) held in front at a speed of more than one metre per second and then stretched forward again a bit shortly before reaching the small flying insect. The left female (in 3) has about 10 cm lead. In evolution, competition was an important pacemaker for the development of peak performance. (Three images composited into one.)

▲

A female Copper Demoiselle has caught and is consuming a large insect with only its antennae and legs are still visible.

A Blue Hawker (*Aeshna cyanea*) disassembles a Common Crane Fly (above). The powerful, sharp and toothed mandibles (right) can even cut through massive chitin parts of armoured insects. Dragonflies are aptly named Odonata (= the toothed ones).

A young Black-tailed Skimmer (*Orthetrum cancellatum*) has caught a male Copper Demoiselle in flight and is carrying it away. Black-tailed Skimmers consumed dozens of the Copper Demoiselles at a watercourse in southern France, demonstrating their aerial superiority. ▶ sn.pub/07qb00

DRINKING

Not only is eating part of the dragonfly's metabolic needs, but equally important is water intake as seen with these Demoiselles drinking (both illustrations from left to right). They plunge headfirst into the water, stay in contact for a short moment and then push themselves up again, the water running off the non-wettable wings. The start is easily achieved with a wide swing.

▼

G. Rüppell, D. Hilfert-Rüppell, *Dragonfly Behavior*, https://doi.org/10.1007/978-3-662-70234-5_16

CLEANING

Given the demanding requirements of flight, it is not surprising that dragonflies spend a lot of time and care cleaning their sensory organs, especially their eyes, but also their wings and then their entire body.

This male Darter (*Sympetrum*) turns his head widely while cleaning the upper side of the right compound eye.

This Banded Demoiselle male (*Calopteryx splendens*) has perched on a Water Lily flower and unintentionally dusted itself with pollen. This is bothersome, and the Demoiselle tries to remove the pollen with its legs.

Intense grooming by a Blue Hawker (*Aeshna cyanea*). The closely spaced grooming bristles of the legs are used like a brush to comb over the eyes and mouth parts. The dragonfly also does this with two legs at the same time (top left). The comb of the cleaning bristles also stands out clearly in colour. (below).

A Spreadwing (*Lestes spp.*) brushes its left eye with its leg bristles. The three ocelli between the antennae are particularly prominent here.

▼

The abdomen is also regularly cleaned (left). In these Demoiselles, this is done with the bristles of the hind legs. Its mobility is trained by extreme bending (stretching exercises) (right). ▼

Once the abdomen is free of dirt, it can be used as a rasp to clean the wings. To do this, it is passed through the folded wings with sweeping movements. ▼

INTER-SPECIES ENCOUNTERS

Dragonflies of different species often encounter each other. This rarely ends peacefully. A male Azure Bluet (*Coenagrion puella*) attacks a male Common Bluetail (*Ischnura elegans*) because it obviously wants to perch in the same place. ▼

Perches in prime positions are highly desirable places sought out by many dragonflies and are often at the centre of disputes. Here a Four-spotted Chaser (*Libellula quadrimaculata*) (right) and a White-faced Darter (*Leucorrhinia*) are fighting for this place.

The larger species try to catch smaller ones. At a small garden pond, for example, a Blue Hawker (*Aeshna cyanea*) frequently hunted darter tandems and in the south of France large numbers of Black-tailed Skimmers (*Orthetrum cancellatum*) regularly preyed on Demoiselles.

▶

© The Author(s), under exclusive license to Springer-Verlag GmbH, DE, part of Springer Nature 2024
G. Rüppell, D. Hilfert-Rüppell, *Dragonfly Behavior*,
https://doi.org/10.1007/978-3-662-70234-5_18

On the Oker River north of Braunschweig, Banded Demoiselles (*Calopteryx splendens*) and Blue Featherlegs (*Platycnemis pennipes*) live together in large density. Both species require water plants as egg-laying sites or perches. This creates problems. ▶

Here there is a lack of space: The Blue Featherleg tandems stubbornly lay eggs in places that Banded Demoiselles dispute. On the left a Banded Demoiselle male attacks an egg-laying White-legged damselfly tandem, which is simultaneously harassed by a conspecific. In the right picture, all places are occupied by White-legged Damselflies, so that the Banded Demoiselle must fly past without being able to land. ▼

The male Demoiselles are looking for a perch in their territory, while the Blue Featherlegs want to lay eggs here. The Demoiselles threaten with their wings and a male pushes with his legs (above).

On another flower, the Blue Featherlegs have almost won thanks to their tenacity and superior numbers (below).

▼

Dominant behaviour of Demoiselles

Here we see conflict resolution by the Demoiselles. The males try to win back the perches where the Featherlegs lay their eggs by pulling the tandems out of the water (top, from left to right) or carrying them away (bottom, also from left to right). Only after carrying the Featherlegs for about 25 cm does the male Demoiselle release the tandem. ▼

FIGHTING

Hardly a minute goes by in which male dragonflies do not fight – against other males or to acquire females. When fencing with their legs, they demonstrate the highest level of agility. In this fight, one of the two male Blue Emperors (*Anax imperator*) has turned on its back to defend itself with its legs. But that is no problem for them: dragonflies are built so that they can passively turn themselves back over through the wings and the long abdomen. But how is the free-moving head supposed to survive the expected impact? Stanislav Gorb in his bionics research group has solved this paradox: dragonflies have a so-called head-arrester system. Chitin clasps with cushion-like protrusions are pressed against the head from the thorax and thus stabilize it during mechanical stresses. This protects the sensitive head attachment from damage during fights or when catching prey. ▲

This male Blue Emperor enters into a fight as if with a battle shield stretched out in front of it – all legs stretched forward. It has probably been through many fights, because all three of its right legs are damaged.

◀

In this fight between two Blue Emperor males (below) the head-arrester system is also engaged. The outstretched legs are extended towards the opponent. In addition, each male's mouth is open for a brief moment (lower male, bottom left drawing) and then closed again to bite. The attacked male (bottom right) flees with a quick turn in an inclined posture. ▼

Two Blue Emperor males race towards each other (from left to right). On impact, all legs are involved in the fight (right, four images composited into one). Mostly this is enough for one male to win the right to stay at this pond for hours. Sometimes with similarly strong and motivated males, the outcome is only made after many minutes of fighting. The winner will then patrol the same pond for several hours, but may also move to neighbouring ponds in search of females. Only when none appear after a period of time patrolling does the territory holder leave the laboriously conquered pond and search elsewhere. Many other dragonflies such as Blue Hawker (*Aeshna cyanea*), Black-tailed Skimmers (*Orthetrum cancellatum*) or Four-spotted Chasers (*Libellula quadrimaculata*) also show such temporary territorial behaviour. ▼

Showdown in Texas: A tandem of the Common Green Darner (*Anax junius*) flies to lay eggs (picture sequence from top to bottom). The males are marked with numbers. The tandem male (1) is separated by bites from an attacker (2, 2nd phase), then it turns around (3rd phase) and bites the new tandem male (4th phase and single enlargement next to it) – without success. It then flies off immediately.

The newly formed tandem is now attacked by another male (3, bottom, left) and falls into the water with a somersault (bottom row, center). Both dragonflies in the tandem drown and are subsequently eaten by water beetles.

▼ ▶ sn.pub/dbd5nd

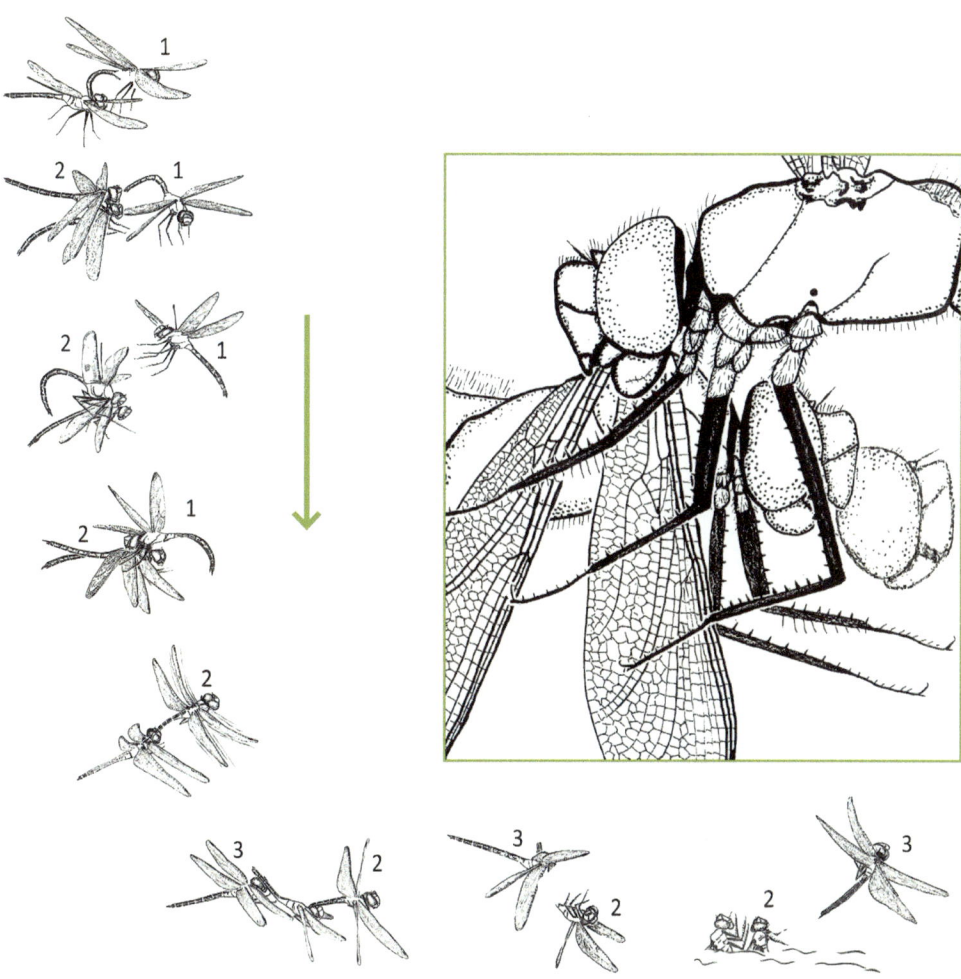

In addition to the clawed and bristly legs (above), the jaws are another effective weapon with which dragonflies inflict injuries on their opponents (below). Here, male Banded Demoiselles (*Calopteryx splendens*) are biting into enemy wings.

▼

Many injured male Western Demoiselles (*Calopteryx xanthostoma*) were observed in the south of France. They had been fighting continuously one early summer on the Céle River. Flooding had led to such a strong current (top) that the

egg-laying females could not submerge and were always reachable for the males who fought fiercely over them. Over 50 % of the males had bite injuries on their wings. Schematic pictures extracted from film sequences.

▼

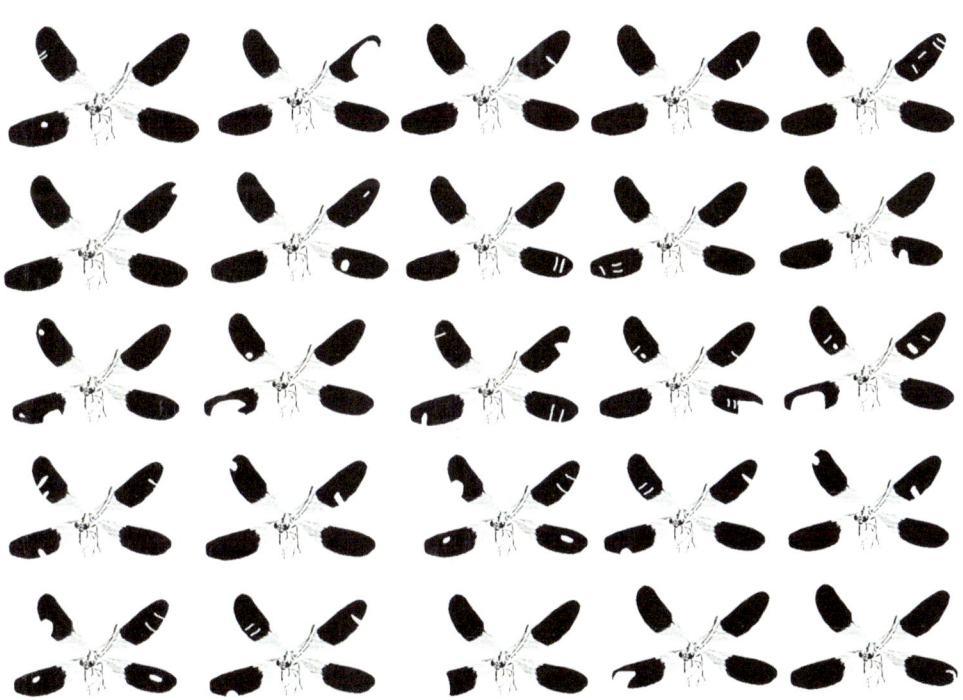

▲

Bite wounds have almost completely destroyed the left hind wing of this male of the Banded Demoiselle (above). Its flight ability is greatly reduced.

The outer part of the injured wing of this Banded Demoiselle male flutters like a flag when flapping. This handicap will make this male a loser, especially when catching prey, during threatening flights, fights and courtship displays. To compensate for the loss of wing area, males with severely damaged wings have to flap around 20% faster and forego the otherwise frequent gliding with wings folded backwards. Both consume more energy and also reduce communication possibilities.

▼

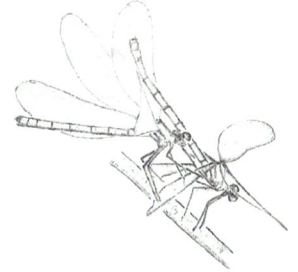

Female Banded Demoiselles also fight for perches by ramming, as seen here. They use a special threatening flight, which is similar to the courtship flight of the males with opposing beats of the front and hind wings.

These female Copper Demoiselles (*Calopteryx haemorrhoidalis*) are chased away by the male. Now the priority for the male is to find a good perch for catching prey . Courtship and mating will come later. ▶

Skimmer fights

Most of the time, the unprotected abdomen is the target for attack, as seen in these male Black-tailed Skimmers (*Orthetrum cancellatum*, image sequence from left to right). The goal seems to be to throw the opponent off the flight path. The attacked male reacts by bending its abdomen to free itself (right).

▼

▲ At the edge of a pond the clearly audible clashing of wings reveals the battle of these male Black-tailed Skimmers competing for females (image sequence from left to right). In the process, the males ram each other and hit with their legs. The lightning-fast adjustment of the wings for flight control takes place within milliseconds.

Two Black-tailed Skimmer males clash while trying to seize a female, but are prevented from doing so by their own fighting: First, three males arrive in front of a female resting in the plant thicket (1), 0.55 seconds later two males attack the female (2) and begin to fight 0.14 seconds later (3). The female is able to escape to the left and the males move away to the upper right (4, another 0.07 seconds later) having lost the female. (Four film images assembled to match time and space.) ▼

These Downy Emerald males (*Cordulia aenea*) also fight for territories where they expect to find females. Only the short-term exposure with an exposure time of 1/3000th of a second reveals the details. Even in their fights, the legs play a big role and are stretched out towards the opponent. Despite risky flight acrobatics, crashes are very rare. ▶

An optimal habitat of the Small Redeye (*Erythromma viridulum*). The males sit on the floating leaves of the pond lilies and often fly up to attack other males or females or tandems.

▼

In the attacks on other males, these Small Redeyes also fence with their legs at lightning speed. The battle shield position formed by their legs is clearly visible whenever they approach other males. In addition, differences in wing movement on the two sides of the body and the usually horizontal head position are visible. This results in agile flight manoeuvres.

THREATENING

Male dragonflies can avoid fights and become pacifists – but only if they have coloured wings. Dragonflies like these Chasers of the genus *Neurothemis* from Thailand show off their coloured wings. This is how they let their conspecifics know how fit they are. The division of territories and thus also mating opportunities can be regulated in this way without fighting.

G. Rüppell, D. Hilfert-Rüppell, *Dragonfly Behavior*,
https://doi.org/10.1007/978-3-662-70234-5_20

▲ Many dragonfly species threaten each other with special modes of flight during which the coloured wings are often held still for short moments. The wings of the Demoiselle *Mnais nawai* in Japan are red with orange (top left), those of the males of the Dark Red Bishop (*Neurothemis fluctuans,* bottom left) hanging in the air in front of each other are reddish-brown and those of the Metalwing Demoiselle (*Neurobasis chinensis*), both from Thailand, are silvery to shiny green (bottom right). André Günther has filmed many jewel-wings such as the *Aristocypha fenestrella* (top right) in Southeast Asia, exotically coloured species with bizarre-looking threatening flights.

▲ A male Banded Demoiselle (*Calopteryx splendens*) impresses here by display-ing its blue wing ornaments. The territorial males often use wing flashing by opening and closing their wings while sitting, and thus demonstrate their ownership and readiness to mate. On beautiful summer days, it sometimes flashes blue in many places along a river.

▲
If a brownish or greenish coloured female Demoiselle (left) sits somewhere nearby in the bank vegetation, it is difficult for the males to discover her. She can calmly watch the flight show of the males and then choose one.

Wavelike or rocking flight paths enhance the conspicuousness of the menacing flights of rival male Banded Demoiselles. Especially in the morning, the air above natural streams is full of dancing males with sometimes slow, conspicuous flight patterns but also interspersed with rapid pursuits in the so-called escalated fights. ▼ ▶ sn.pub/3s1367

Frontal threatening is particularly effective in male Banded Demoiselles: the blue wing surfaces appear particularly large and imposing, while the spread out wings block the opponent's path. Whoever can hold out the longest wins and gets the disputed territory and thus access to any arriving females. An even stronger threatening display against rival males is achieved by males of the Beautiful Demoiselle (*Calopteryx virgo*) and Copper Demoiselle (*C. haemorrhoidalis*) where they replace the normal threatening flight with the mating flight style comprising parallel, high-frequency wing beats. ▶

Water spraying as a reinforcing threat signal

A discovery on a small tributary of the Céle River in the south of France (Lot department): Here, two Beautiful Demoiselle (*Calopteryx virgo*) males are competing for females in a good spot. They circled each other many times with the usual flapping of all four wings. However, the threatening fight is intensified here by using the water surface!

The first three photographs are from the same sequence. You can see how the left male flaps its right hind wing on the water (2), creating circular water waves (3). Now the second male also flaps its right hind wing on the water (3). Both males create these water circles in the direction of a rival and directly in front of a female sitting on the bank. In picture 4, taken a few minutes later, the lower male touches the water surface twice with one wing and also creates water splashes with his legs (left in the picture).

Such an amplification of the threat signal could only be observed here in many hundreds of hours of observation of Demoiselles elsewhere in Europe.

Getting a Female

The males of damselflies and dragonflies always outnumber females at the water's edge. They are waiting for females like these males of Common Bluets (*Enallagma cyathigerum*). ▼

When a female Common Bluet emerges from laying eggs under water, many rival males pounce on her. There is high competition between males. ▼

G. Rüppell, D. Hilfert-Rüppell, *Dragonfly Behavior*, https://doi.org/10.1007/978-3-662-70234-5_21

Males search for females everywhere. Here a female Blue Emperor is seized by a male Lesser Emperor (*Anax parthenope*) while laying eggs. ▶

Life on water lily pads
Males of the Small Redeye (*Erythromma viridulum*) also pounce on any females they see, even if they are already in tandem (left).

With its newly achieved conquest (bottom) a male flies to less contested places to mate undisturbed. ▼

FEMALE DEFENCE AGAINST MALES

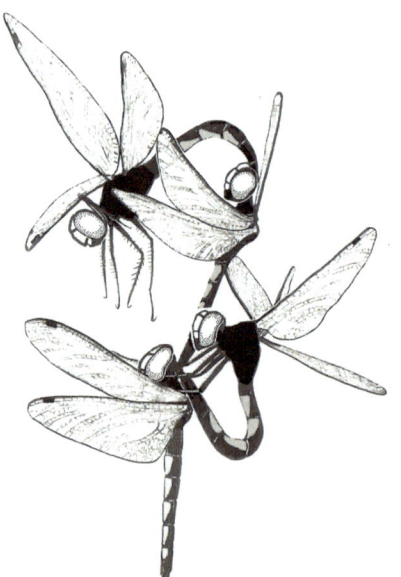

Dragonfly females are often exposed to the persistent advances of males – they will actively defend themselves against this. Here, three males have attached themselves to a female of the Ruby Whiteface (*Leucorrhinia rubicunda*) – all without success. This quartet dissolved again without having accomplished anything.

◄

Flying in tandem

There is great competition between males for females and they will try anything to mate. Many Damselflies, and also some Dragonflies, secure a female in tandem and then fly around together to mate and lay eggs. Despite this bond, rival males will still attack. Here a female Azure Damselfly (*Coenagrion puella*) bends her abdomen forward, reducing the landing area for the incoming male.

▼

G. Rüppell, D. Hilfert-Rüppell, *Dragonfly Behavior*,
https://doi.org/10.1007/978-3-662-70234-5_22

▲

Sharp turns

Female Blue Emperors (*Anax imperator*) often lay their eggs in floating stems where they are visible to males from afar. No wonder they have developed special defence techniques. They fend off intrusive males with a unique evasive flight manoeuvres. They perform a dance-like pirouette manoeuvre.

If a male approaches, they swing their abdomen forward and, with special wing beats, enter a fast, tight circular motion that the male cannot follow. Top photo: One female lays eggs, while the other (left) notices the approaching male (right) and then outmanoeuvres him by flying a pirouette with her abdomen swung forward, and the male flies by without further engagement (bottom photo).

The female Blue Emperor (right) also uses her outstretched legs to defend herself during the tight turn (above)..

The tight turn can be performed immediately after take-off (below) or in flight (right), sometimes several times in succession.

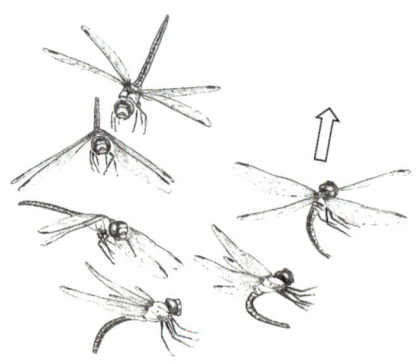

The females fly in a very tight curve radius, which the males cannot manage to follow. ▶

▼

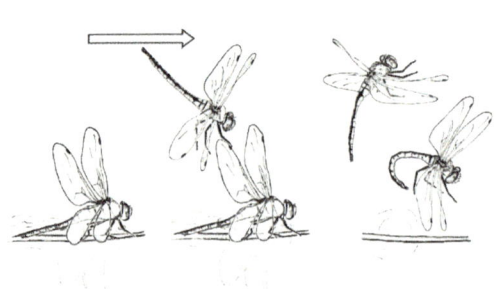

Hold On

Blue Hawkers (*Aeshna cyanea*) often appear at garden ponds. Unlike Emperor Dragonflies, the females lay their eggs in marginal plants, moss cushions, or damp soil. Males find them hard to spot here because of their camouflage colouration. But when a male has found a female and has formed a tandem, the female holds on to the substrate (right).

This usually makes the suitor give up quickly. Measurements have shown that the female can develop ten times more holding power than the male has pulling power.

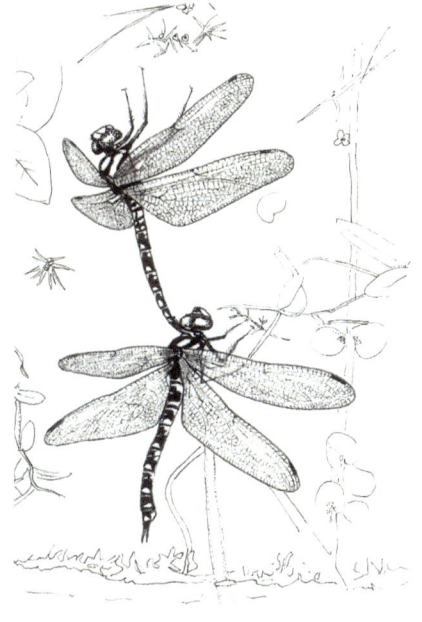

87

Playing dead

Some Blue Hawker females simply drop and remain motionless when pursued by a male (drop and stop). Here a male has discovered a female and is trying to drag her out of the vegetation – but without success. The female has escaped and the male is sitting there alone. (bottom, right)

▼

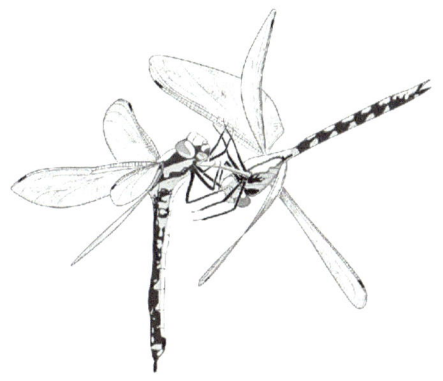

The decisive moment: This female of the Blue Hawker forcefully pushes a male back with her legs – biting probably also plays a role in the short confrontation.

Females camouflage themselves with colouration similar to males. Males of the Common Bluetail (*Ischnura elegans*) sometimes hold their females in tandem for many hours (up to 7). This means a lot of wasted time for the females. As a result, they have developed male-like colourations; seen here on the left pairing wheel. Male-coloured females probably experience less disturbance because they are mistaken for males, and not females, by other males.

In high density populations, a female Black-tailed Skimmer (*Orthetrum cancellatum*) defends herself while laying eggs in a sitting position against an attacking male by raising her forelegs (left) and wings. When it threatens to tip over, she quickly puts her forelegs back on the ground (right).

Whenever foreign males want to attach, the females of Ruby Whiteface protect the area behind their heads, by raising their legs (arrows). On the right, a female in the mating wheel directs her legs towards the attacking male, just like the female and male of a tandem on the left.

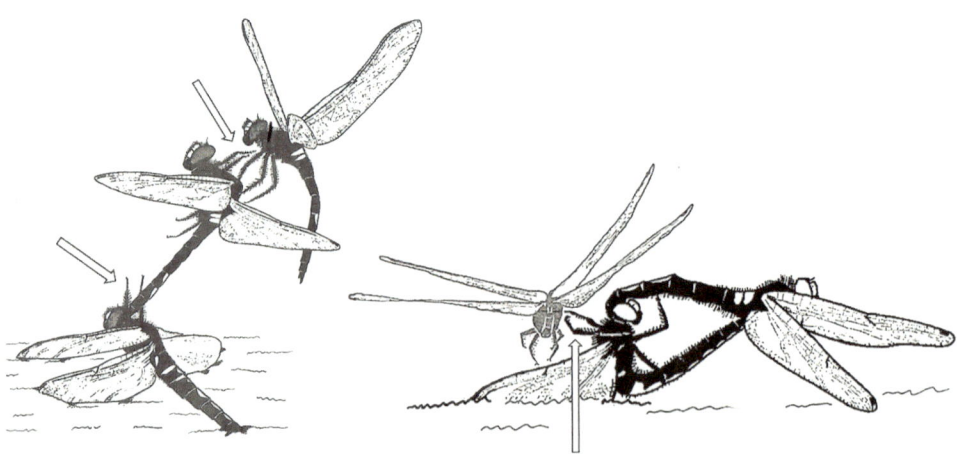

This watercourse (right) with adjacent still waters (left) offers a wide diversity of bank and plant structures and thus good development opportunities for still water and flowing water species. Also important are the higher resting places and hiding places in the trees during the night, in bad weather, or for hardening after the maiden flight. ▼

Passive resistance is chosen by this Copper Demoiselle (*Calopteryx haemorrhoidalis*) female when refusing the male's mating attempt. It stiffens and bends its abdomen forward, but does not couple with the male. This refusal lasts several minutes until the male finally flies away.

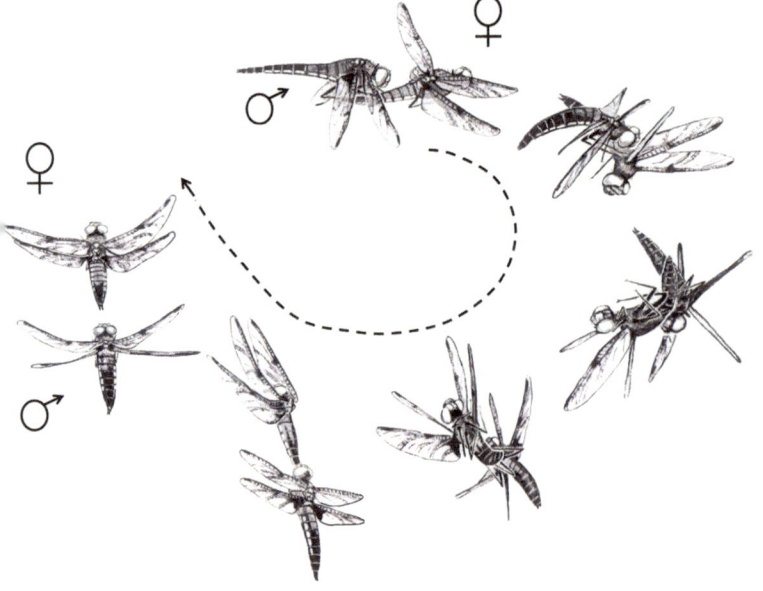

◀ **Loopings**

At a small garden pond (10 x 4 m), 3–8 males of the Four-spotted Chaser (*Libellula quadrimaculata*) were present. Arriving females were immediately pursued by them and they tried to escape the attentions of these males. Here the pursuing male can already grab the female (at the beginning of the picture sequence at the top centre). However, she can shake off the attacker by flying a loop. Another time, a female uses a male's ramming kick (below) to fly a loop and escapes. ▶

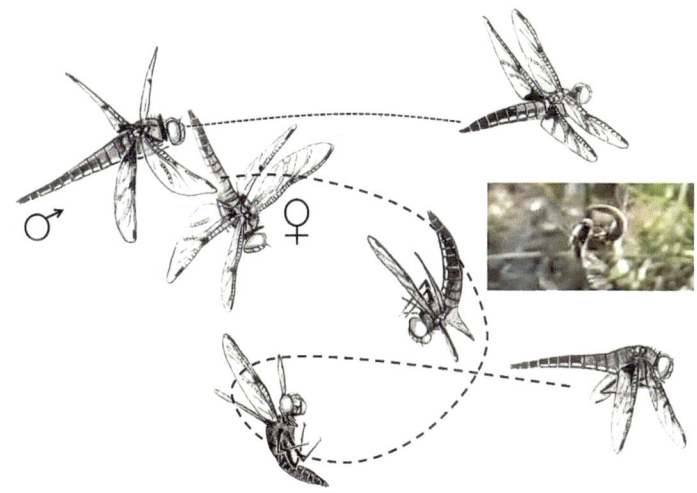

▼ **Fatal attack by the Four-spotted Chaser**

An egg-laying female is seized by a male (starting on the left, black arrow and photo). The female climbs onto the attacker in flight and bites him. The male tries to shake off the female (large drawing). Suddenly the male falls over into the water and remains lifeless. A few seconds later it is captured by a frog (bottom centre). The female then continues to lay eggs in the immediate vicinity

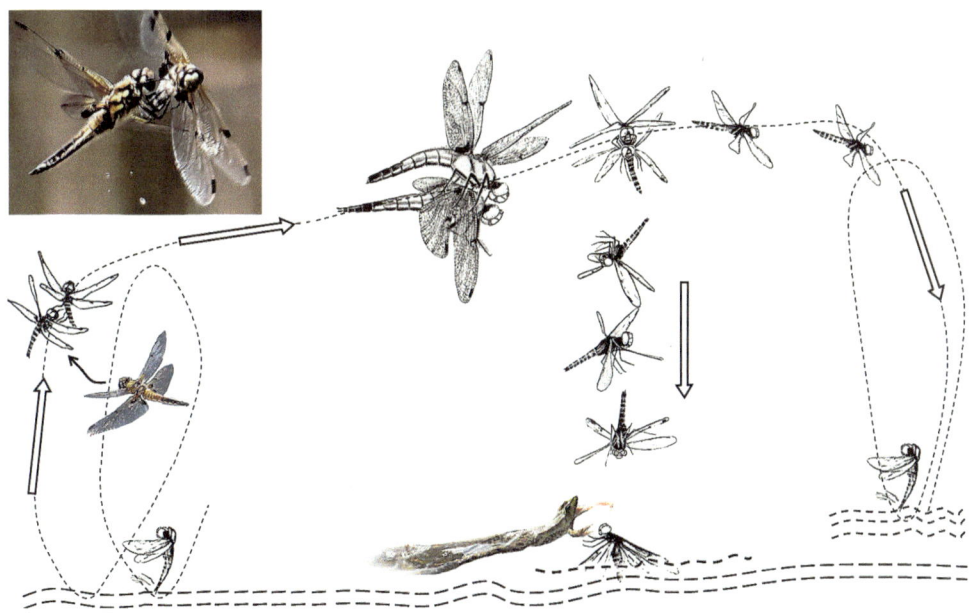

(right). It is possible that during this unexpected attack by the female, the male's head-arrester system was not active and the nervous system collapsed. Although this is only a single observation, such behaviour is certainly not unique among dragonflies in Europe.

Failed male defence

Typical mating of the Blue Hawker with the cooperation of the female (top row). The female actively bends her abdomen towards the male's secondary copulatory organ and holds on to his abdomen. This is quite different in the case of mating forced by the male (bottom row). Here the female stiffens up and does not grasp the male's abdomen.

▼

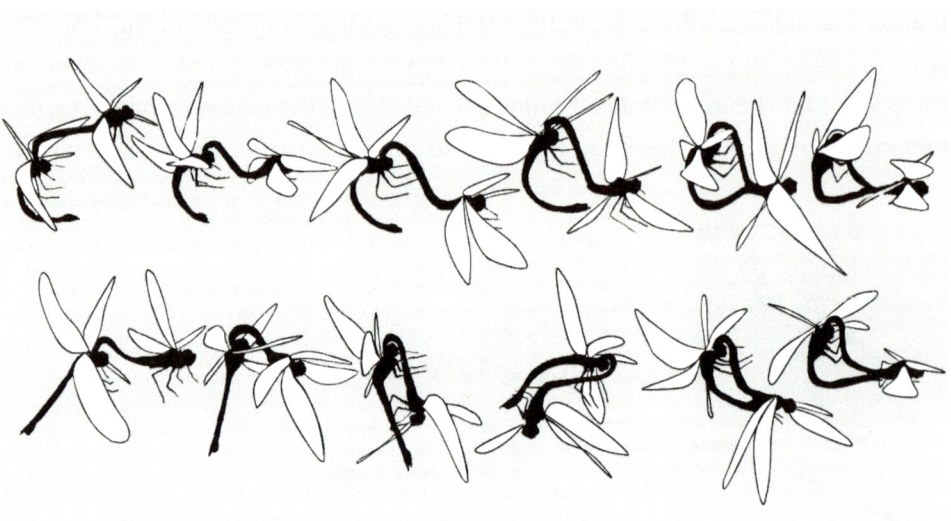

This male has grabbed a female while she is laying eggs (beginning at 1), lifts her up (2, 3) and presses her intensely against him. The female shows no co-operation, compared to typical mating, which can also be seen from her non-grasping legs and stiff body posture (4). By strongly flexing his abdomen (5), the male finally succeeds in reaching a mating position, but the female still does not cling to him (6). ▼

This female Blue Hawker does not cooperate either, stiffens up and does not fly and they both crash. Still on the water, the male tries to reach the mating wheel (bottom right). Sometimes they manage to grab a plant, more rarely to fly up again. But dragonflies can swim.

▼

▲ A Blue Hawker female swims by slow rowing movements using her wings at a frequency of 1–2 beats per second and at an angle of 40–45 degrees until it reached the shore (from left to right).

Female Common Bluetails lay their eggs alone. They avoid mating, which in this species can take many hours. This female fell into the water while fleeing from a male (1). Then it begins to groom itself (2). To do this, it scrapes the wing surfaces clean with its abdomen (3). ▼

Libellulid flying acrobats

Darters (*Sympetrum*) like Chasers (*Libellula*) have very good manoeuvrability due to their short bodies and high flapping frequencies. Although this male has tried to escape while in tandem, a rival has grabbed the female and is trying to bite him to take her over.

He doesn't succeed, but it's still a super-fast and highly precise manoeuvre. The tandem pair managed to avoid the attack, demonstrating the durability of the tandem's connection.

▼

Courtship

Male dragonflies are not always forced to chase and catch females. In species with coloured wings, like the Demoiselles, they can dispense with aggressive threatening and conquest battles and use displays of their coloured wings to perform courtship flights in front of the females. Then a female chooses a male quite peacefully (from left to right). Females will sometimes watch for a long time, because the energy-intensive courtship flight is a test of the male's fitness. The males now flap their wings very differently than during normal flight. They increase the flapping frequency to 50 beats per second (more than double that of normal flight) and no longer move their front and rear wings simultaneously, but out of phase. In the Banded Demoiselle (*Calopteryx splendens*), they move in exactly the opposite direction. All species of Demoiselles do this slightly differently. This is how the females recognize the males of their own species and also determine how fit a suitor is. It is now clear why Demoiselles need such special wings with many longitudinal veins – with the high flapping frequencies and the many reversals of flap direction during courtship flight, they must remain stiff and as visible as possible over their entire surface. The many small wing cells also help to carry out these rapid strokes. ▼

G. Rüppell, D. Hilfert-Rüppell, *Dragonfly Behavior*,
https://doi.org/10.1007/978-3-662-70234-5_23

Female Demoiselles (seen in the background) sometimes back away from a male and allow it to continue displaying if they are not yet convinced of its fitness. It took a male Banded Demoiselle almost 800 beats to be accepted by a female for mating. In the middle, you can clearly see how the male's wings now move in opposite directions during courtship flight, i.e. the front wings move backwards and the rear wings move forwards at the same time. The other damselflies are egg-laying Featherlegs. (Montage of three individual images of males from the same period and the same location). ▼

Flight changeover within milliseconds

A female appears and the Banded Demoiselle male (right), previously flying upwards in parallel flight, immediately switches to high-frequency courtship flight (left). Dashed line: path of the left forewing tip.

▼

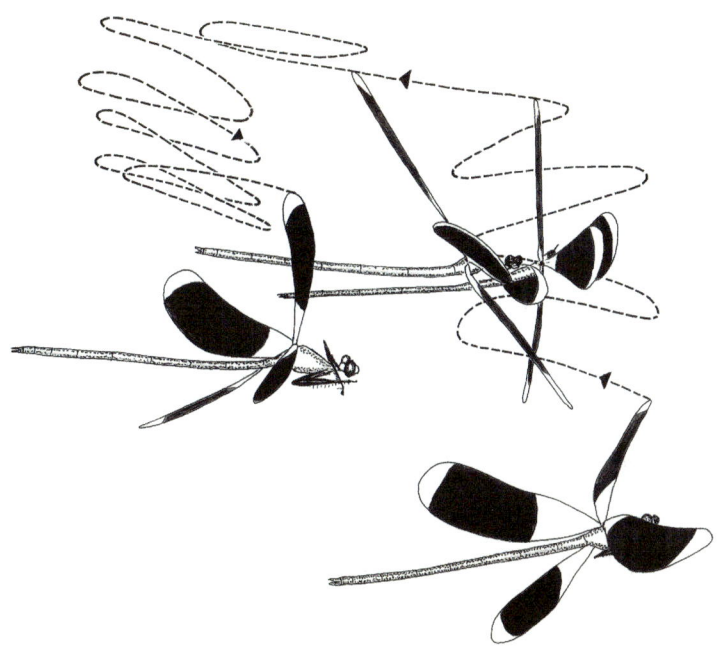

During courtship, male Demoiselles display the end of their upturned abdomen as a sexual signal, their so-called 'brightly glowing lantern'.

◄ They often land on the water's surface (left, middle and bottom) and drift a bit downstream. This demonstrates to the female that the flow speed of the water is ideal for egg laying and larval development. The larvae of Demoiselles need flowing aerated water to breathe sufficiently and to ensure a good supply of suitable prey. In addition, the males seem to impress the females with this risky behaviour (avoiding attacks from frogs and fish). (Here you see Copper and Banded Demoiselles using this sexual signalling)

▶ sn.pub/2jzamg

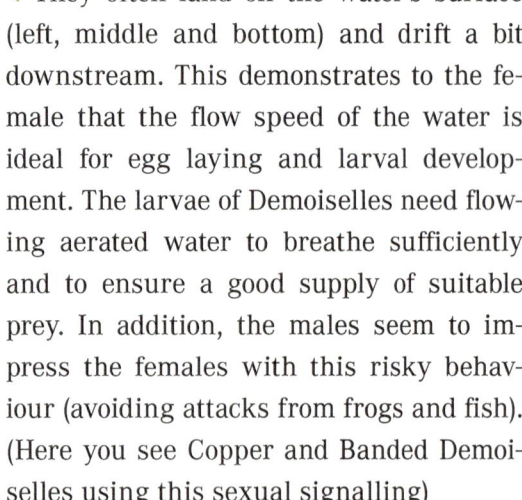

MATING

▲ Here the blue male and emerald green female of the Banded Demoiselle (*Calopteryx splendens*) are mating. Mating is the most important behavioural element in reproduction. The mating of dragonflies takes place in a unique way, the so-called dragonfly wheel, in which the two partners connect together and can even fly. The wheel can only come about because the males have an additional copulatory organ on the front abdomen, the so called 'secondary genitalia'. This is where the male will transfer his sperm to from the sexual organ at the rear end before mating can commence. Once this is done the female in tandem (as shown in the photograph above), bends her sexual opening to make connection with the male's secondary genitalia and receive his sperm. But that's not all. What happens in the bodies of the damselflies and dragonflies during mating has led to a scientific sensation.

G. Rüppell, D. Hilfert-Rüppell, *Dragonfly Behavior*, https://doi.org/10.1007/978-3-662-70234-5_24

◀ In order to be accepted by the female for mating, the male Banded Demoiselle must approach the female very slowly during courtship flight (1).

He is often rejected by the female flapping its wings or even flying up. Only if he displays the high-frequency courtship flight for long enough and its wing ornaments are correct will the female allow him to mate (2, 3). Both then remain in the pairing position for several minutes until the sperm is transferred (4).

A peaceful body contact is established: at many points the two Copper Demoiselles (*Calopteryx haemorrhoidalis*) touch each other and send out positive signals, so that both partners continue the behavioural steps towards mating. ▼

A male Demoiselle attaches itself to a female. The male's appendages fit exactly onto the female's prothorax. This is (arrow) how the tandem is formed (two photos of the same process composited into one). ▼

▲ In this unique position (right), the male Banded Demoiselle transfers his sperm from his sexual opening at the rear end into the copulatory organ at the front of the abdomen. The female will then bend her abdomen and connect her sex opening to this during the formation of the mating wheel.

▲ Here too, in the Azure Bluet (*Coenagrion puella*), sperm filling in the secondary copulatory organ takes place in exactly the same way (from left to right).

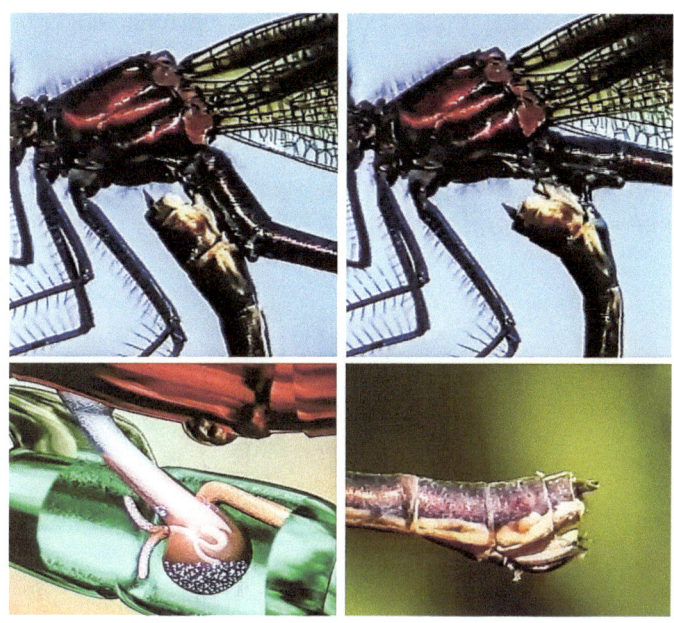

Why Demoiselles became so famous

The discovery of the unique process of dragonfly mating by Jonathan Waage went viral around the scientific world: He had found that male Demoiselles will remove the sperm of any previous matings from the female during the initial phase of mating and then replace it with their own. This sperm replacement is shown here in full swing with the Copper Demoiselle:

The male (top left) presses his secondary copulation organ deep into the female. There, he probes with two long appendages into every corner of the female sex organ, in order to remove as much as possible all the sperm of a predecessor (schematic drawing bottom left). To do this, he pulls his copulation organ upwards in a movement of his body and brings the foreign sperm to the surface (top right). This is repeated several times until he quickly gives his own sperm to the female.

The whole process takes about two and a half minutes. Afterwards, the female remains sitting for a while and scrapes the removed sperm with genital appendages aside (bottom right) before flying to lay eggs with the new sperm. The male will guard his female well there, so that the same thing does not happen to him as happened to his predecessor.

107

Obviously, early mating success is even more successful for male damselflies. The earlier a Demoiselle male can mate in the morning, the more mates it can have throughout the day and ensure that his genes are passed on to the next generation. Here the male Banded Demoiselle flies back to the water after mating. He will lure "his" female there and while she lays her eggs he will guard her from rivals and also from predators. ▼

◄
This male of a Banded Demoiselle even presses with his legs on the female – an invitation to lay eggs.

The peaceful behaviour of the Banded Demoiselles at low density

black wings = male, grey wings = female

wu	warming up
pc	prey catching
of	observing females
t	threatening
c	courting
m	mating
sw	settling on water
gel	guarded egg laying on water

Unsuccessful mating attempt

This male Banded Demoiselle has made a mistake. Driven by a high hormone level he has mistaken this young and not yet fully coloured male for a female and is now trying to mate with him. Another interpretation suggests that such behaviour aims to fixate a rival. Held in this way, he could not approach the female.

Two pairs of the Blue Chaser (*Libellula fulva*) and a lurking male of the same species. Mating with sperm exchange is now taking place here. Just landing on the overhanging reed stalk in the wheel position is an aeronautical masterpiece. ▶

Last in – first out (sequence of images from left to right)
When there are many males competing for a few females, as with these Four-spotted Chasers (*Libellula quad-rimaculata*), it has to be quick. Then there is no time for a prolonged exchange of sperm. The male grabs the female in the air as quickly as possible (1, 2) and mates in flight (3).

This only takes a few seconds. In this case, his sperm is only transferred to the female, but the foreign sperm from a predecessor is left there untouched. The partners then separate (4), and the female goes straight to lay her eggs. In this "last in – first out" behaviour, the eggs are inseminated with sperm from the male that last mated with the female.

▼

Royal mating

Blue Emperors (*Anax imperator*) often mate while hanging from a branch or plant stalk, because lifting the approximately one gram heavy female while sitting is difficult. Here, the foreign sperm is also replaced by the new male's own, which can take up to thirty minutes with these large dragonflies. The male here is missing almost half of its left hind wing. Nevertheless, it was still able to capture a female.

Fast communication during the Black-tailed Skimmer mating

While sitting, this male Black-tailed Skimmers (*Orthetrum cancellatum*) has transferred his sperm. Now the male signals his intention to release the mating wheel by briefly trembling his wings (above). Then it bends its abdomen upwards and opens the anal appendages (left). At the same time, the female folds its legs from lying behind the head to face upwards and then pushes them towards the male.

The mating is over, and the partners separate. The male flies back to the water, the female stays perched on the ground for a while.

Here a pair of Black-tailed Skimmers show off their flying skills. It flew eleven turns in eight seconds to avoid attacks from rival males (top right) and then find a landing spot to finish mating. The start of the flight is at the black dot. The small pictures are spatially assigned to the pairs' flight path.

▼

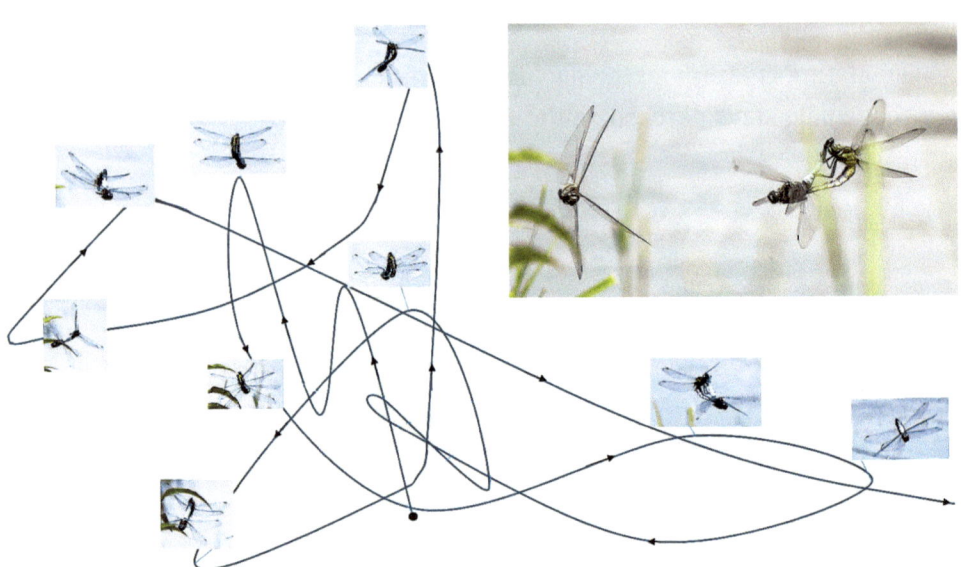

Mysterious looping

Here something surprising occurs during mating: A male Black-tailed Skimmer has caught an egg-laying female and flies forward with her (starting from the top from left to right, then bottom row from right to left). It performs a looping in a fifth of a second. Fluid sprays from the female (4th phase). Whether this is water or sperm from the predecessor or already fertilized eggs is unclear. We have tried in vain to collect this fluid. The male then replenishes his sperm supply in the secondary copulatory apparatus (bottom row from right to left, 1st phase) and mating takes place (bottom row, left) before the female detaches after a few seconds and flies off again to lay her eggs (far left). Black-tailed Skimmer

or Four-spotted Chaser males at high male-density repeatedly seize the egg-laying female and mate with her several times in succession, so that there is a high probability that their sperm will fertilize the eggs.
▼

Dragonfly mating wheels are quite stable formations. This pair of Yellow-spotted Whiteface (*Leucorrhinia pectoralis*) was attacked by a male (left), but was able to defend itself, so that the rival was unsuccessful. ▼

115

ALTERNATIVE REPRODUCTION

When a period of fine weather breaks out in summer after a time of cool weather, hundreds of Banded Demoiselles (*Calopteryx splendens*) emerge and clouds of them now flutter on the rivers. Then their behaviour changes dramatically. None of the males has room and time to establish a territory to defend or per-

G. Rüppell, D. Hilfert-Rüppell, *Dragonfly Behavior*,
https://doi.org/10.1007/978-3-662-70234-5_25

form courtship flights. Now it's just about: chasing females. These are pursued intensively. In this picture, there are about 20 males to be seen on just a little more than 5 meters of riverbank. Females are also there, but hidden on the bankside.

Now there is a crowd in the river. The males of the Banded Demoiselles can still threaten each other but defending a perch or even a territory is pointless – there are too many rival males there. Now it's about being the fastest in the pursuit of females. ▶

Several Demoiselle males besiege floating water plants. Two others fight in the bottom right for a female. Now the females can no longer choose their mates by their visual characteristics – only an indirect choice is still possible by flying away and being caught by particularly fast males. ▼

117

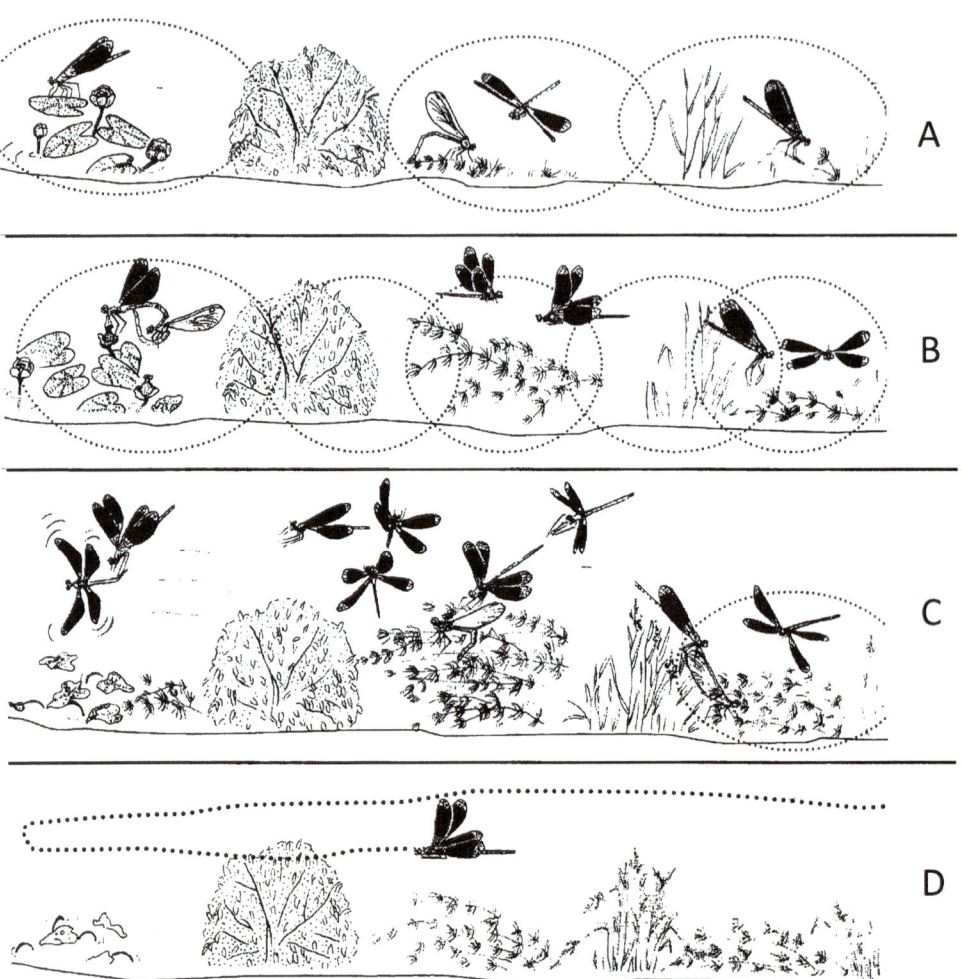

▲ Thus, the behaviour of Banded Demoiselles changes over the course of a year: With the emergence of the first water plants, the first Demoiselles also appear in good weather in May (at A). They establish large, isolated territories which they defend, and then court females.

If the favourable weather continues, more and more dragonflies emerge, the number of territories becomes more frequent and smaller (at B). The conflicts between rival males increases until the density becomes so great that there is no time left for territorial behaviour. Now aggressive, alternative reproductive behaviour spreads (at C). Finally, in late summer, only isolated male Demoiselles fly often long distances in search of females (at D).

In alternative reproductive behaviour, males of Banded Demoiselles are in the majority, while females are rarely found. The males now chase after every female, even if they are already coupled to a male in tandem. ▼

Females try to escape the male hunt by diving. This female did not quite manage it. It is seized by a male and carried away for renewed mating (bottom right). The male runs a high risk of being preyed upon. ▼

What a transformation: previously peacefully courting at low density, now aggressively fighting. A male of the Western Demoiselle (*Calopteryx xanthostoma*) attacks a tandem (1). It bites the opponent in the wing (2) and drags him away from the female (3).

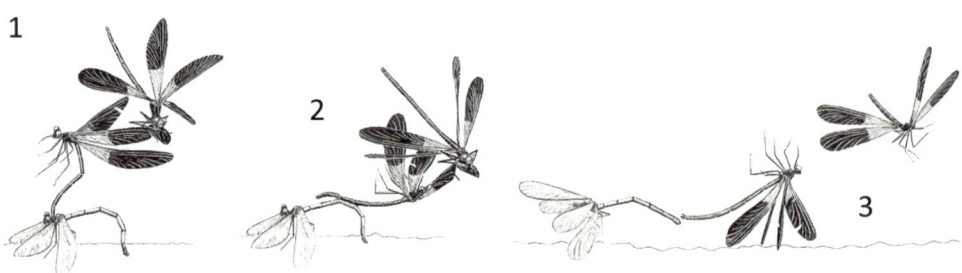

In the heat of battle, sometimes two male Banded Demoiselles also couple to a female, either one after the other as in the left picture or both together at the head of the female as below. In both cases, mating did not occur, but the three separated after intense tugging.

◄ ▼

120

Tandem separation

To get females, every means seems to be right. The attacking male Banded Demoiselle bites a tandem male in the abdomen, where the female is coupled (above), until this male releases the female.

Immediately, the new male couples (below). Whether the female has now got a stronger or fitter male is not certain. The males can hardly defend themselves against biting at the rear end. They can only try to shake off the attacker. ▶

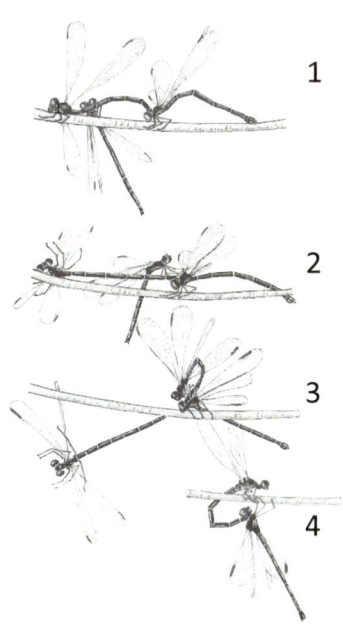

◀ Tandem separation also in Western Willow Spreadwing (*Chalcolestes viridis*). Here an attacker bites the male of the tandem (1), runs to the female (2), bites the male again (not shown), which then lets go (3). The new male then takes over the female (4).

121

▲ In softwoods like alders or willows, the Western Willow Spreadwing (*Chalcolestes viridis*) drill holes into the bark and lay eggs there communally.

This female Banded Demoiselle (*Calopteryx splendens*) fights free from an attacking male. It had just landed on the water surface to submerge when a male pounced on it. But the female was on her back (left) and thanks to the non-wettable wings, she could take off again and fend off the attacker with her legs and fly away (middle). He is left behind (top right, three images composited into one).

▼

▲ These Banded Demoiselle males also go empty-handed. They have discovered a female emerging from laying eggs and try to couple at the same time (left). But they get into fighting so much that they detach from the female and flutter away. No one gets to mate here.

Why males usually lose races with females (sequence of images from left to right)

Here, too, several male Banded Demoiselles are chasing a female (left and centre, outlined in white). They begin to threaten each other and all rush past the female towards another displaying male so that the female then escapes. They then threaten each other again in the air (right). ▼

Threatening distracts

This mutual threatening during the female pursuit can also be recognized here by the fact that the males simultaneously hold their wings in recoil position for a while. While they glide forward, they also send threat signals to the rivals – and this has slowed them down, so that the female escaped.

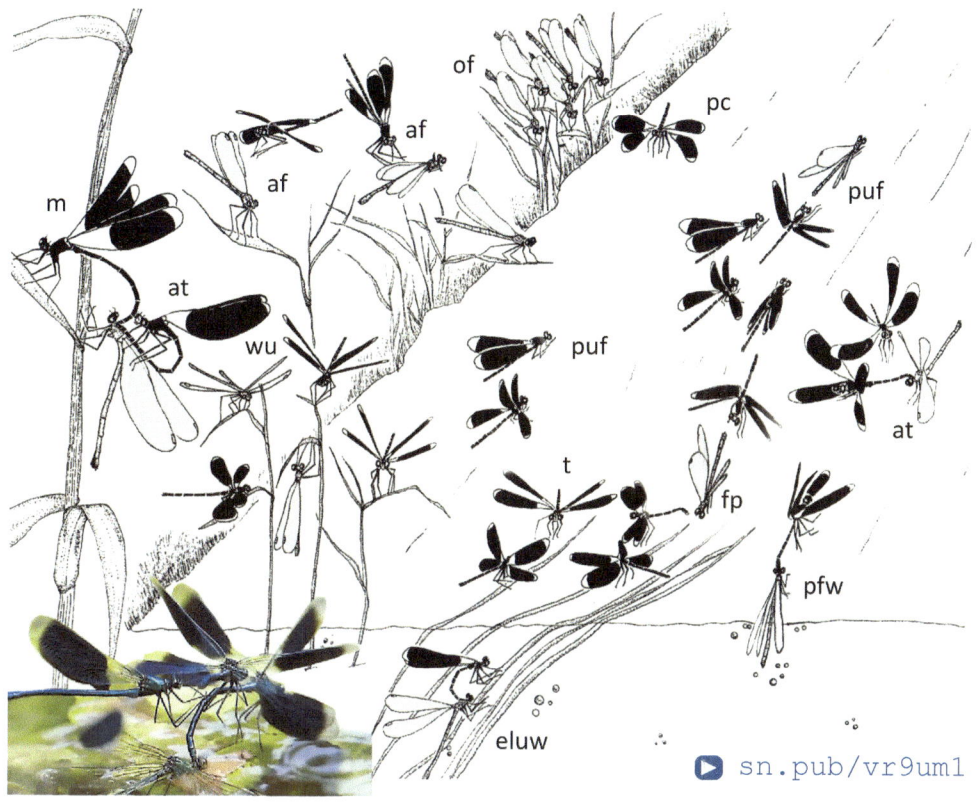

sn.pub/vr9um1

Aggressive, alternative reproductive behaviour of Banded Demoiselles at high densities

black wings = male, grey wings = female

pc   prey catching

t   threatening

wu   warming up

of   observing females

m   mating

eluw   egg laying under water, male diving, too

fp   females plunge diving

puf   pursuing females

af   attacking females

at   attacking tandems

pfw   pulling females out of water

125

Egg Laying

Rivers of low mountain ranges are home to Common Goldenrings (*Cordulegaster boltonii*). The males will often patrol along long distances of the river in search of females (bottom left).

The female Common Goldenring flies up and down in suitable places, laying eggs in plant masses (right). Egg laying is a crucial moment in reproductive behaviour, but it is dangerous: in all bodies of water, predators such as frogs, fish and birds are waiting for the egg-laying dragonflies. ▼

G. Rüppell, D. Hilfert-Rüppell, *Dragonfly Behavior*, https://doi.org/10.1007/978-3-662-70234-5_26

▲ Egg laying in tandem

The male's sperm only fertilises the females eggs when they are laid. They slide down the female's fallopian tube to the exit of the reproductive organ where they receive a dose of sperm from a sperm receptacle. Only after the eggs are laid can the male be sure of becoming the father of the offspring. For this reason, many males of damselflies, but also some of the dragonflies, hold the females in tandem until the eggs are laid – including the Darters (*Sympetrum, see montage*). This is how the male protects "his" female from rivals. Laying eggs in tandem while in flight has further advantages: The eggs are well distributed.

In addition, the pairs are only briefly in the dangerous hunting range of predators due to the short immersion.

The female dragonfly releases dozens of eggs with each dip of her abdomen. Dipping in with pinpoint accuracy requires the highest level of agility and manoeuvring skill. Only a few downward flights come to nothing. ▶

127

Damselflies species of Darters (*Sympetrum*) lay eggs in groups. This significantly reduces the risk of being preyed upon by birds or frogs. As tandems cannot fly as agilely the group protection is a great advantage. ▼

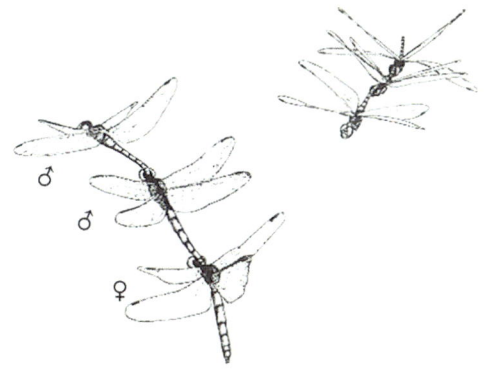

With Darters, the male usually leads the female in tandem to lay eggs. To do this, it must coordinate its flight, so that the tip of its partner's abdomen touches the water surface with each dipping action. Here in Hokkaido in Japan, however, another male has attached itself to the tandem male. Surprisingly, the leading, third male also flies so precisely that the female hits the water exactly, even though it had not mated with the female.

The males of the dragonfly genus *Tramea* (studied in Texas) detach from the female near the potentially dangerous water surface – but then only very briefly (image sequence from right to left). Immediately after egg-laying they catch the female again. The females, however, also have an advantage from this unique behaviour as they can escape when it suits them.

Many damselflies such as these Azure Bluets (*Coenagrion puella*) lay their eggs directly in plants while perching on them. They gather in groups to do this.

When clouds cover the sun, the tandems leave the water and land on bankside plants (left) and only lay eggs again in sunshine.

▲

Some dragonflies also as this Blue Emperor (*Anax imperator*) lay their eggs directly in plant parts. The eggs are protected there so that they produce far fewer eggs than species that freely lay in the water.

On floating stalks, they have a clear view so that they can detect approaching frogs or birds at an early stage and escape predation.

They hardly pay attention to damselflies, they even let them land on them (right).

▶

The life insurance of egg-laying dragonflies like this Blue Hawker (*Aeshna cyanea*) are their outstretched wings. Here she holds them still for a moment – but usually the female dragonflies vibrate them to keep the flight muscles warm and ready for take-off. ▼

After mating, the male guards "his" female (here *Calopteryx splendens*). Unguarded females also steal in to benefit from the guarding. The male, who has usually only mated with one of them, here is driving away a rival.

▼

In areas with a high density of males, female Banded Demoiselles lay their eggs underwater. This female has just managed to do so, a male arrives too late. ▶

▲

Egg-laying underwater

A female Banded Demoiselle submerges to lay her eggs in underwater plants (top). To do this, she saws a hole in the aquatic plant with saw-like cutting flaps (bottom left; bottom right rear end of a female) and pushes an egg into the substrate. The females inhale air, which flows along the diffusion gradient from the water into the silvery air envelope that surrounds them – a "physical gill". They can survive under water for over an hour. Dozens of eggs can be laid during the diving time. Two dangers must also be survived there: Fish and chilling. Especially when the water is still cold early in the flying season, the females are unable to emerge after surfacing and drift down the river to an uncertain fate.

134

In most cases, however, they manage to float up after the dive, as with the Copper Demoiselle (*Calopteryx haemorrhoidalis,* right) in Southern France. The female floats motionless to the water surface and flies off with a few wing beats. ▶

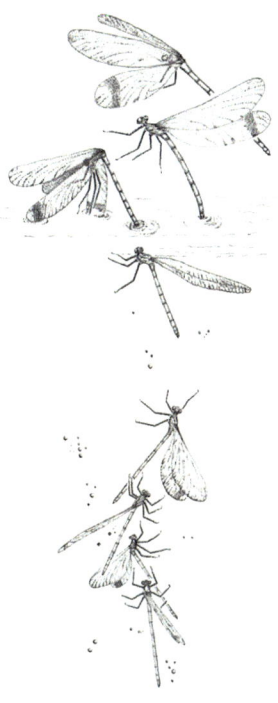

If there are not many males around, female Black-tailed Skimmer (*Orthetrum cancellatum*) lay eggs during continuous flight. Even then they may be chased by males. Therefore, they often only come to the water at times when the males are less present, such as during cloudiness or early or late in the day. ▼

135

How females trick males

If there are many male Black-tailed Skimmers chasing them, the females try to lay their eggs as undetected as possible. Two tactics help the females: one (above) is to sit motionless and let the eggs swell from the genital opening until a ball of eggs has formed (top left). Then they fly up briefly and rub off the ball of eggs in the water (bottom left). Immediately afterwards, they sit down again and let the eggs swell out again without moving (bottom right, the swelling process begins again). ▼

The other tactic of Black-tailed Skimmer females is to lay eggs while sitting still for a long time. Only for small changes of location do the females fly up briefly and then immediately continue laying. With both tactics, the females are only visible to the males for a fifth of the time due to their short flight and thus avoid many chases and risky pairings.

▶

When laying eggs in a resting position, the females press the ball consisting of dozens of eggs onto the surface (starting top left). Then they bend their abdomen upwards and detach the ball of eggs, which is now attracted by the water, and it sinks (bottom), until it has completely disappeared in the water (bottom right). If the ground substrate is too dry, this rubbing off does not always succeed and the females need several attempts until successful laying. ▼

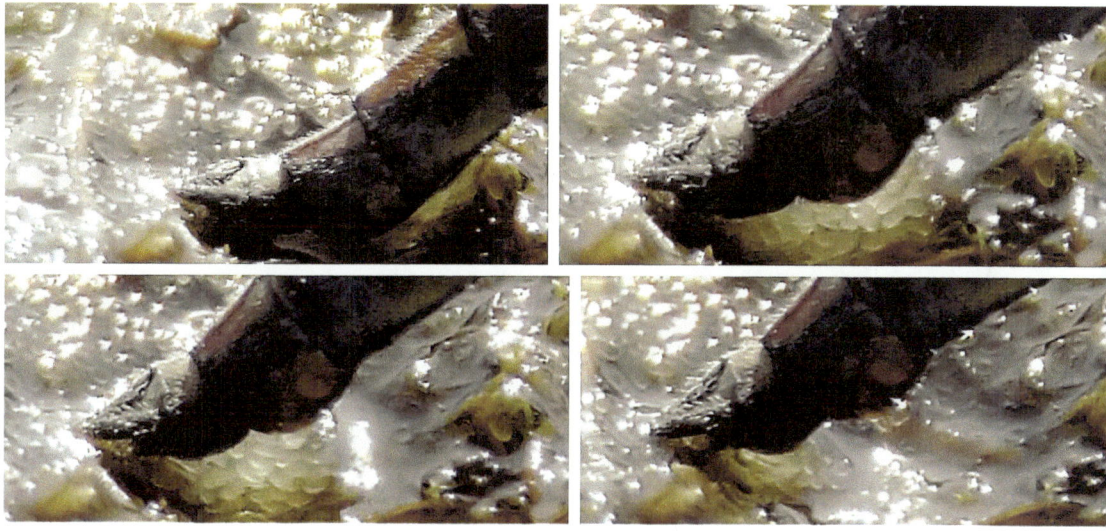

▶ sn.pub/nof80k

LARVAE

▲

Dragonfly larvae live in water. In this way, the larva and imago develop in two different feeding environments and do not compete with each other. In order to grow, larvae go through a series of stages or stadia and have to moult each time. The number of larval stages is between 8 and 17 depending on species but can also vary by five moulting stages within a species. Depending on latitude, altitude and temperature regime, several generations of dragonflies may hatch each year or it may take several years for the next generation to appear. In fishponds, as above, dragonflies have a hard time. The fish feed on them and so the larvae show an adapted behaviour in that they perform all movements more slowly and also take longer to develop. The larvae ensure the success of the flying adult dragonflies by feeding intensively. The heavier a larva becomes, the larger it becomes as a flying insect.

The original version of the chapter has been revised. A correction to this chapter can be found at https://doi.org/10.1007/978-3-662-70234-5_33.

Dragonfly eggs are just millimetres in size, and the larvae in the first stages of development are only slightly larger. Left: early stage of a dragonfly larva, right: young damselfly larvae already fighting over a perch by flapping their tail fans (caudal appendages). ▼

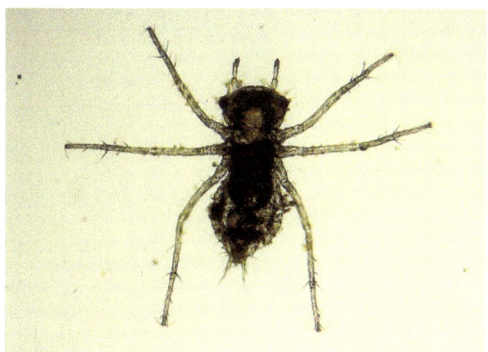 

Damselfly larvae have three caudal appendages, which they use to move around using swimming movement like oars. They also breathe via these large exchange surfaces, which are traversed by breathing tubes (tracheae). If the larvae are caught by predators on their tail fans, these can fall off like the tail of a hunted lizard and the larva remains alive. ▼

▲ This larva of a Blue Hawker has just shed its skin for the last stage before emerging (above). Its increase in size compared to the previous empty larval skin is clearly visible. But all body structures are still pale and soft for a while – a dangerous condition when they are vulnerable to predation.

▼

The back end of a dragonfly larva (*Aeshna*) is armed with five sharp spines (left), which it will use to defend itself against attackers. The larva breathes (arrows above right) by sucking in water by rhythmically inflating the abdomen and opening the anal opening (bottom left) which is then expelled again after closing the anus (bottom right when closing).

If you look inside the rectum at the rear end (right, double focus image), you can see the respiratory organs, the tracheal gills, as thread-like projections. They are arranged in six double rows and absorb oxygen from the water.

▼

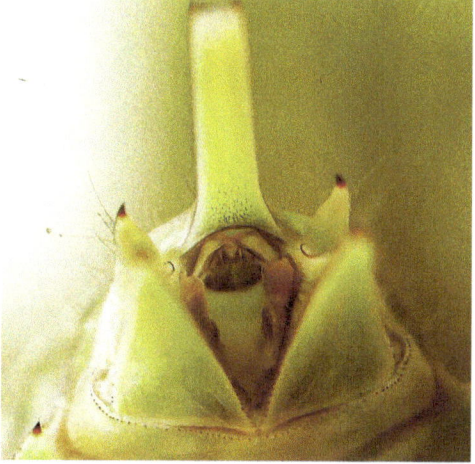

▲ Sometimes a whitish growth can be seen on the dragonfly larvae. These are usually single-celled organisms, which are carried along by the dragonfly larvae on their journey. Among them, Vorticella species (top picture) can be found on Hawker larvae (left), but these commensal organisms do not cause any harm to the dragonfly larvae.

The larva of the Four-spotted Chaser (*Libellula quadrimaculata*) is covered with hairs. This breaks up the body outline so it can remain well camouflaged. ▼

Dragonfly larvae catch their prey using their labial mask (labium) which is armed with two sharp hooks at the front. This unique adaptation has evolved from the lower lip of dragonfly ancestors and is deadly efficient. Above you can see the labium of a Blue Emperor in its folded state.

Here, a larva of a Blue Hawker (*Aeshna cyanea*) is ejecting its labial mask at a prey that it narrowly misses (bottom right). ▼

◀ A larva of a Blue Hawker strikes with the labial mask to catch a mosquito larva. The mask is tensioned by muscles and fluid pressure and then shoots forward. The catching hooks fold together horizontally – probably an adaptation to hunting in open water. The mask is unique in the animal kingdom and offers two advantages: being able to lurk motionless and then eject the mask, which increases the range in which it can catch its prey. ▶ sn.pub/srzjzb

Fast backstroke swimming

If dragonfly larvae like this Blue Hawker larva needs to escape a predator it quickly expels water from its rectum. This generates a recoil that propels the larva forward quickly.

A type of jet propulsion! Here you can see the generated water jet which shoots out of the water with some plant parts at the top. During forward movement the legs are held close to the body (right) resulting in less resistance. It thus reaches a speed of 30–40 cm per second. ▼

▲ A live larva (top) and an empty larval skin (exuvia) of a Common Goldenring (*Cordulegaster boltonii*). The dragonfly has hatched from the opening at the top of the exuvia. The huge, toothed spoon-like mask covers the front of the head. The vertically arranged "teeth" are probably adaptations to the lack of food in fast-flowing streams. The development period of 3–5 years is correspondingly long.

145

The larva of the Common Goldenring often sits buried in the sand and silt at the bottom of the stream. Sometimes only the eyes look out (top). When a freshwater shrimp approaches (2nd picture from top), it suddenly appears on the surface of the sand and then shoots out its mask surprisingly quickly (3rd picture from top). Its limbs are very flexible in relation to each other so that it can dart out in many directions. At the bottom it releases an entangled grain of sand. ▼

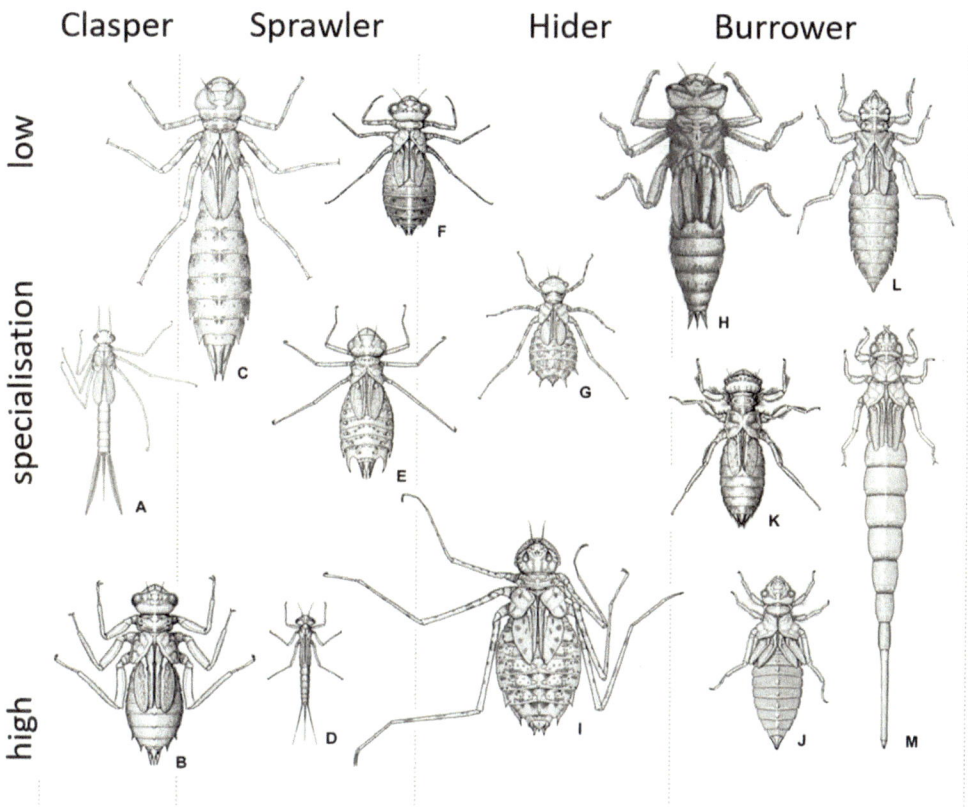

| Clasper | Sprawler | Hider | Burrower |

low

specialisation

high

▲

Body shapes and action types of various dragonfly larvae according to F. Suh-ling et al (2015), artwork by O. Müller; species on the dividing lines are interme-diate forms, occurrence: without designation Europe, otherwise indicated. A *Calopteryx* , B *Zygonyx* (Africa, Asia), C *Anax*, D *Ischnura*, E *Pantala*, F *Diplaco-des* (Africa, Asia), G *Brachythemis* (Africa, southern Europe), H *Cordulegaster*, I *Phyllomacromia* (Africa), J *Onychogomphus*, K *Orthetrum*, L *Notogomphus* (Af-rica), M *Neurogomphus* (Africa). These body shapes contain a lot of information about their way of life. Good backstroke swimmers such as *Anax* (C) have a slender body shape. Many running water species (H, L, M) also have a slender body shape and short, broad legs when they live buried. But there are also spe-cies such as *Phyllomacromia* (I), which live in running waters in calm zones and have correspondingly long legs. They stretch them out sideways for stabiliza-tion, but probably also to locate prey. The latter may also apply to Demoiselles.

Hidden species often have a flattened body that does not take up so much space (G, I). Many species have spines on their abdomen for defence (B, E, G, I), which the deeply buried larvae do not have. The degree of specialisation means that the larvae are less or more closely tied to the specified activities.

▲

By collecting the larval skins (exuviae), the colonisation of a body of water can be determined. This is part of the exuviae of various dragonflies collected from a small pond and proof of them breeding there. The number of emerging adults is often very high – however, the number that survive emergence and fly off successfully is sometimes much lower, as the newly emerged adults (known as tenerals) are a feast for predators. In addition, poor weather can sometimes throw a spanner in the works and damage emerging adults.

PREDATION

▲ Not all dragonfly larvae hatch. Many are eaten by predators – even the eggs: these mallard ducklings are particularly dependent on protein-rich food at the beginning of their lives and look for anything that tastes good, including dragonfly eggs in aquatic plants.

▲

Especially water plants such as the Milfoil (*Myriophyllum*), in which Demoiselles, often guarded by the male (top right), had laid eggs close together from above, are literally chewed through by mallards.

▶

© The Author(s), under exclusive license to Springer-Verlag GmbH, DE, part of Springer Nature 2024
G. Rüppell, D. Hilfert-Rüppell, *Dragonfly Behavior*,
https://doi.org/10.1007/978-3-662-70234-5_28

▲ Dragonflies are actively hunted by frogs (Edible Frog *Pelophylax esculentus*). This is worthwhile for these predators, because a Blue Emperor brings the frog a valuable source of energy (around 1 gram of biomass) with one successful leap. If it was to catch mosquitoes instead of these large dragonflies, it would have to jump several hundred times to get a similar amount of food.

The frog's tongue is an effective catching organ that significantly extends the reach for catching insects, such as adult dragonflies. It reminds of the labial mask of dragonfly larvae, which also extends their reach so they can hunt many frog tadpoles. Thus through evolution, the catching tools have been extended in both these adversaries, frog and dragonfly.

The slime on the frog's tongue only becomes really sticky at the last moment when it shoots out in the air. This prevents the frog from catching itself.

▼

Hunting frogs jump at an average speed of 1.5 meters per second – in addition, the flinging tongue also speeds towards the prey at a speed of 4–6 meters per second (above ground) at its tip.

▲ How dangerous the water surface is! – here shown by this chilled, crashed damselfly. It has no chance to escape as it sticks to the meniscus of the water and afterward to the frog's tongue.

A frog jumping from behind can catch a tandem pair of Azure Bluet (*Coenagrion puella*) (starting top left, each to the right). The danger was difficult for the guarding tandem male to recognize due to the view against the sun. Also, the jump took less than a tenth of a second.

The frog detects everything that moves with its eyes piercing through the surface of the water.

153

Escaping sideways

A frog jumps at an egg-laying female Blue Emperor (*Anax imperator*, left). It's a matter of milliseconds and millimetres. Only by lightning-fast, sideways acceleration does she manage to escape (bottom, left).

The instantaneous uplift from the water, even from a back position, is effortless for the dragonfly, because its wings are non-wetting and the frog cannot re-orientate its jump quickly enough to catch it (bottom right).

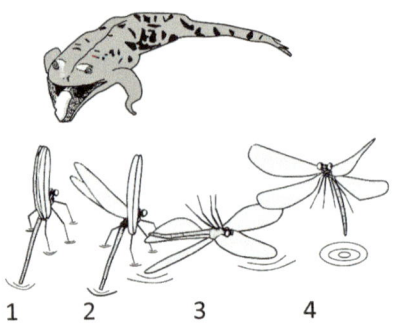

1   2   3   4

Dragonflies attempt to get out of the way of the frog's jump and the strike of its catapult tongue using acrobatic flight manoeuvres. Sideways tipping, like that of the egg-laying female of the Demoiselle (left 1–4), or a rapid double beat of all four wings (right) helps a dragonfly to get out of the line of fire of the frog (dark wings = downstroke, lighter wings = upstroke).

This female Blue Emperor also saves itself with a great acceleration: (starting top left then to the right). It achieves this through unusual, wide and fast wing beating. Unlike normal flight, the wings are swung so far that they almost touch at the top (2nd picture).

They also beat upwards now, not phase-shifted and nearly together (1st picture). This provides more thrust and is also rarely used, because it requires a lot of force. But here the largest European dragonfly saves itself. It can shake off the frog's brief touching, sticky tongue.

▼

The rescue stroke (arrow) of the Blue Emperor during the frog attack: Both pairs of wings were simultaneously struck back upwards at a high speed and steep angles. As a result, the dragonfly shot forward and the frog into the void. (black and grey wings: successive phases of the fore and hind wings)

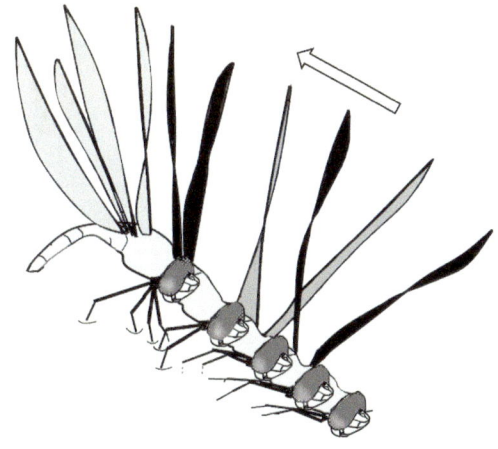

Hit

A frog has sneaked up on an egg-laying Migrant Hawker (*Aeshna mixta*) and shoots his tongue at her in the jump (starting top left to right, bottom row the same). Although only a tip of the tongue hits the wing, it sticks (3rd phase), and the frog can pull the large dragonfly down to its mouth (below).

But this was also only possible because the dragonfly was just starting at this moment to fly off and was no longer holding on to the plant stem.

▼

▲

A frog jumps (from left to right) after a tandem pair of Darters (*Sympetrum*). First phase: The frog aims exactly where the dragonflies are right now. In the second phase, the frog has already jumped 3 cm upwards, but the tandem has only moved a little sideways. By the time the tongue is fully extended (right), the tandem pair has already flown more than 8 centimetres sideways and escapes.

A frog (above) is lurking in the water for a Four-spotted Chaser (*Libellula quadrimaculata*, in the middle of the picture) and a White-faced Darter (*Leucorrhinia*, above). In normal flight, dragonflies escape effortlessly, but in the bottom picture a male was pre-occupied in a fight with another male. This cost him his life.

Frogs calculate their success rate before they jump. During one jump, a frog sneaked up to a patch of algae, where despite being further away to the prey, it was able to push off better – with success! ▼

▲
Caught

A Large Red Damselfly (*Pyrrhosoma nymphula*) is swallowed. To do this, the frog stuffs it into his mouth with his left front foot (bottom).

159

A frog aims at two Azure Bluet (*Coenagrion puella*) tandems (from top to bottom). However, it shoots its tongue right between the two tandems into the void (2), so that both pairs can escape sideways. The left tandem, which had previously just sat there, is already much further away from the frog than the one on the right, which had previously laid eggs. The group of damselflies laying eggs behind it on the right also flew away afterwards. ▶

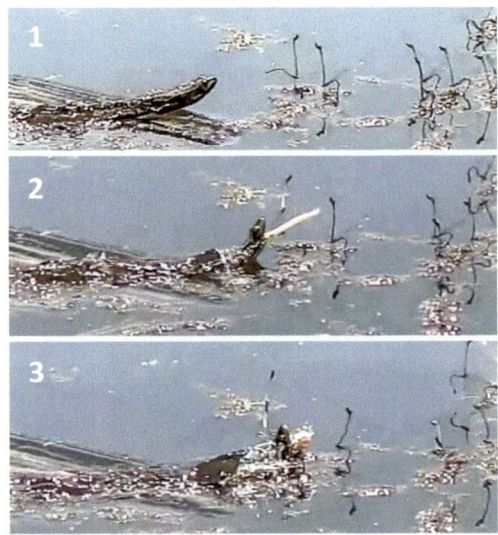

The guarding tandem males of these egg-laying Azure Bluets (from top to bottom) stand almost vertically (white bars indicate the orientation of the tandem males) as a hunting frog jumps towards them (1). Some of these males quickly lean away to the side from the frog's jump (2,3). Only those directly in front of it remain disoriented and are pushed onto the water by the frog's body and some are swallowed (4).

The directions of damselflies already in flight are shown as dashed lines. The perceived wisdom that the central animals of a group are best protected does not apply here.

◀

▲ Bee-eaters (*Merops apiaster*) can even catch fast dragonflies in flight.

◀ The House Sparrow (*Passer domesticus*, top left), together with conspecifics at a small pond, caught about half of the approximately 200 Large Red Damsels (*Pyrrhosoma nymphula*) hatching here in one year.

▲ This Blue Emperor was only a few minutes away from flying off. It was already vibrating to get warm when the Blackcap (*Sylvia atricapilla*) caught it. This small bird systematically searched the pond and caught several emerging dragonflies to carry back to its young. ▶ sn.pub/5w5mbe

This Blackcap is grabbing a freshly emerged Four-spotted Chaser which is unexpectedly only half a metre in front of our camera. We had waited for the maiden flight of this dragonfly – in vain. ▼

▲

Obviously attracted by the catching activity of some sparrows, this magpie (*Pica pica*) caught about a dozen Large Red Damsels in a short time.

162

Wagtails (*Motacilla alba*) on migration tried to replenish their energy, at an almost dry lake in September – by catching dragonflies. In the midst of a swarm of egg-laying dragonfly tandems, the white wagtail is briefly irritated (left), but still manages to catch a few pairs of dragonflies. ▶ sn.pub/bcb4hd

This was achieved when this tandem flew in search of a suitable egg-laying site (bottom left). Such pairs were in danger and were also preyed upon by the wagtails. However, a few passing dragonflies were too fast for the wagtails, which tried to catch them at around 3 m/s about 2 meters after their take-off (right).

163

▲ **Like dragonflies, birds of prey catch their prey with their legs and feet**
Hobbies (*Falco subbuteo*) catch dragonflies (two pictures mounted to one). The protein-rich biomass makes the effort worthwhile. At a speed of around 10–15 m/s, the hobby races towards the dragonfly patrolling over the water (top). With its left foot, it is able to grasp one of the dragonfly's wings and pulls it along. During another capture, Thomas Plack, who took these pictures, was able to take close-ups of the prey being manipulated and eaten.

▲
◄ It is interesting to note that the hawk bites off the dragonfly's thorny legs and flings it away.

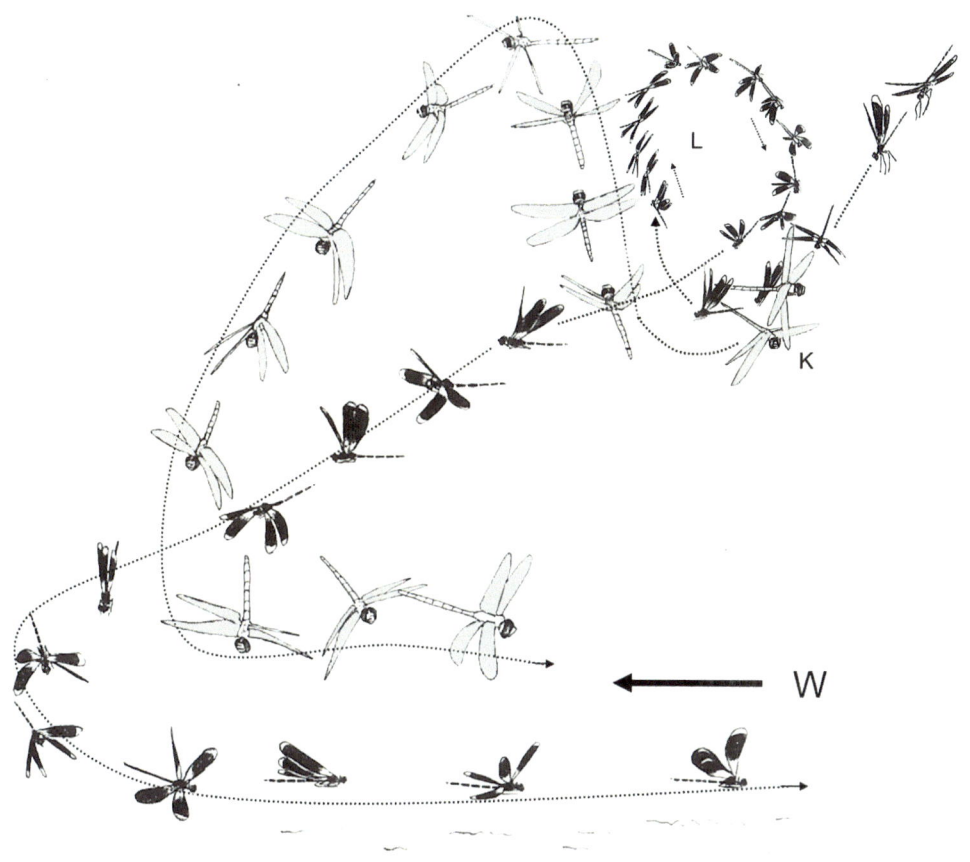

▲

Failure of a hunt by a Blue Emperor

Dragonflies also catch dragonflies and damselflies. Here, the agility of a Banded Demoiselle triumphs over that of a Blue Emperor (starting at the top right). The Blue Emperor narrowly misses the Demoiselle and cannot grasp it (at K). The Demoiselle male then quickly flies a loop (L) and escapes upwards, although the Blue Emperor pursues her.

Afterwards, the Demoiselle flies down close to the water, where the large dragonfly does not follow. Demoiselles are safer from frog attacks there due to their great agility and acceleration ability, than the much heavier Emperor Dragonflies, which avoid these situations as much as possible. (W shows the wind direction).

Blue Emperors (with grey wings) even hunt mating pairs of Black-tailed Skimmer (*Orthetrum cancellatum,* black) like here (image sequence from left to right). This time, however, without success. After a steep curve to the right, the Blue Emperor cannot hold onto the prey. The mass of the Skimmer pair is almost as large as that of their attacker, and the joint work of their eight wings opens up an escape route for the pair. ▼

▲
A male Blue Emperor has caught a Darter (*Sympetrum*) and is eating it.

Death during mating

A male Common Bluetail (*Ischnura elegans*) attacks a tandem of White Feather-legs (*Platycnemis latipes*). It bites into the thorax (left), then separates the front body and begins to eat the abdomen of the female full of eggs (bottom). Common Bluetails are also very successful when it comes to feeding they eat a wide range of different sized prey, including other damselflies of their own size.

With a mighty leap, this backswimmer captured a female Azure Bluet as she was laying eggs. The tandem male was able to escape. Now the damselfly is being sucked out by the bugs proboscis. ▼

Spiders catch many dragonflies, especially damselflies which rarely escape from the webs.

▼

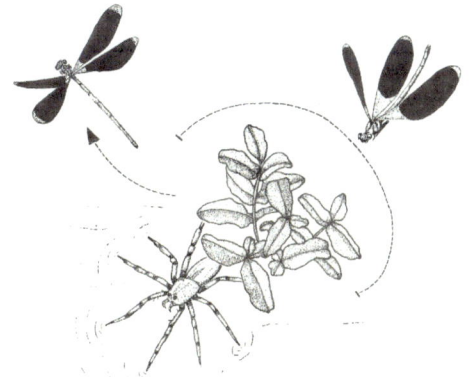

▲ A Banded Demoiselle (above) reacts to the presence of a hunting spider: it avoids the side of the water plant from which the spider had previously performed several attack jumps on it.

▶

Here a Banded Darter (*Sympetrum pedemontanum*) has been caught by the single thread of a spider's web on its evening flight to the roost. Visibility was already poor; it had overlooked the one thread.

169

A Common Bluet (*Enallagma cyathigerum*) has caught a damselfly and then gets caught in a spider's web. The spider (bottom right) immediately begins to spin the captured dragonfly around. The damselfly has lost its prey and flies off shortly afterwards. ▼

Hornets also hunt dragonflies and damselflies. Here, this strong predator approaches the Banded Demoiselles at their sleeping place in the evening. Two Demoiselle males have spread their wings to absorb heat, two others are still in a late mating.

170

Behavioural Diversity of the Demoiselle

▲

The adapted behaviour of the Demoiselles, which they have also shown when being hunted by spiders, is astonishing. It demonstrates that they can learn quickly. It should also give us pause for thought to draw some conclusions about their behavioural performance.

Demoiselles have to show many different, finely tuned behavioural reactions and frequently need to make decisions. This is very different from most other dragonflies. An important underlying reason for this behavioural expansion is probably the acquisition of wing coloration. This has opened up completely new areas of their behavioural repertoire: for example, non-contact fighting through threats and courtship.

When a female approaches a territorial male, many difficult decisions are immediately required of her: does she head for this egg-laying area with this male, does she land or fly on, or does she accept the courtship or not, – these are such necessary decisions.

The males also have to make similar decisions: to continue threatening or to end the threatening flight, which wings to keep still during it? The two front ones or all four, or just one? A male has to decide this in a fraction of a second.

This is all new in the field of dragonfly behaviour. But Demoiselles also master the old, archaic behaviour of the other, clear-winged species, namely when the density of Demoiselles increases significantly. Now they act with active and often aggressive searching for females which include physical fights and risky diving manoeuvres. It is therefore not surprising that Demoiselles exhibit unusual behaviours such as threatening behaviour produced by creating water waves and splashes or carrying away other damselflies.

Thus, the emergence of high coloured blue wings has resulted in a plethora of behaviours that place damselflies at the forefront of behavioural adaptability. It may be that our behavioural observations were more intense in the Demoiselles than in the other dragonflies, but none of the behaviours listed above were seen elsewhere.

172

EMERGENCE AND THE MAIDEN FLIGHT

When they are fully grown, dragonfly larvae come ashore and emerge from their larval skin. This process can take a few hours in unfavourable weather conditions – which is sometimes fatal for the flight candidates.

Here, a male Large Red Damsel (*Pyrrhosoma nymphula*) emerges from its larval skin. The dragonfly increases its body volume by stretching and fluid pressure. Air is also stored in the body of the flying animal so that a low flight weight can be achieved.

▼

Synchronised emergence of many Large Red Damselflies at a small pond

Around 200 larvae emerged from the water here over three days. The increasing day length instinctively indicates to each dragonfly species when it is time to emerge. A series of favourable weather conditions then sometimes cause hundreds of them to leave the water at the same time in just a few hours. In most cases, the larvae can even react to a sudden deterioration in the weather or the appearance of a predator by retreating into the water – but not always; mass emergence losses can also occur. ▼

▲ A Large Red Damselfly hatches. Just pulling the six, delicate legs out of the rigid case of the larval skin undamaged, is precision work. Here, this delicate quick-change artist has become partially dislodged during the process: it is only attached to the stalk with one empty leg sheath of the old larval skin – the others have already detached.

174

What a transformation: the empty larval skin (exuvia) will fade – the Blue Emperor warms up, dries and will soon start a life, full of speed and drama.

In the dark of night, many dragonfly larvae like this Blue Hawker climb up plant stems unseen by predators. This decisive phase is imminent: emergence. The transformation begins now: intestinal respiration ceases, and air is breathed through the tracheal system.

The decisive moment during emergence (left): Does the larval skin burst at the predetermined breaking point so that the head, thorax, legs and wing structures can be easily detached from the old shell? The emergence of this Blue Hawker (right) has almost worked. However, it remains questionable whether everything dries off so favourably that the larval skin falls off here and the dragonfly can still take off. ▼

▶

A classic emergence accident with this dragonfly has led to it getting stuck in its larval skin. This injured dragonfly was certainly utilised by birds.

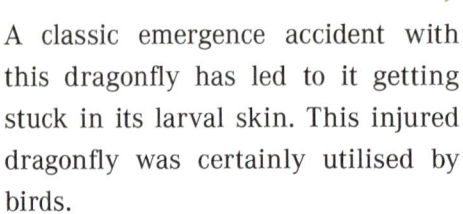

Reducing weight is the top priority after emergence. This newly emerged Downy Emerald (*Cordulia aenea*, above) and the dragonfly in the picture on the right excrete excess body fluid. It can drip from the tip of their abdomens for several minutes.

▼

▲ The wings of this freshly emerged or teneral Blue Hawker are highly reflective and glisten in the light; they are covered with a still fresh, thin layer of wax. The shine is lost after a few hours.

Hundreds of dragonflies emerged at this small pond and many frogs and birds preyed on them. ▼

▶

A Four-spotted Chaser (*Libellula quadrimaculata*, above) has emerged and shows the astonishing increase in volume and structural difference between larva and flying insect.

This Common Clubtail (*Gomphus vulgatissimus,* below) takes its maiden flight straight upwards after a long drying period after emergence.

▼

▲ A Blue Emperor spreads its wings forever after emergence. It will never fold them like this again.

The linings of the breathing tubes (tracheae) are also shed and can be recognised as white threads on the exuvia.

180

Now it has warmed itself by muscle trembling and is just about to make its maiden flight directed high up into the protection of the trees.

▼

▶ sn.pub/ok25bv

DRAGONFLY SWARMS

These Four-spotted Chasers (*Libellula quadrimaculata*) were filmed by M. Koshkin in Kazakhstan when millions of them performed a peculiar spectacle. Groups of them moved towards each other, then flew concentratedly downwards, reminiscent of waterfalls, and then continued flying or landing on the ground. Later, they covered bushes and herbs for roosting (following page). Dragonflies can appear in masses, especially when they form migratory swarms.

The Wandering Dragonfly (*Pantala flavescens)*, for example, crosses the Indian Ocean, while the Common Green Darner (*Anax junius*) migrates hundreds and thousands of kilometres across North America, flying around 60 km southwards every third day and stopping to feed in between. But migratory swarms can also be observed in Central Europe. Large migratory swarms of dragonflies can be seen on coasts, or also on hill chains, in search of new habitats. Despite a comprehensive overview by Michael May, little is known about many aspects of these migrations, such as what triggers them or the energy supply required to achieve them.

A summer day is coming to an end for these Copper Demoiselles (*Calopteryx haemorrhoidalis*). Together, the bright females and some dark males capture the last rays of the day on a branch. Newcomers are warned by threatening gestures that most places are already occupied.

▼

Dew-covered dragonflies glow in the early morning sun in autumn meadows, where they have roosted overnight. Since they are completely inactive at this time, they cool down so much during the night that the water vapour in the early morning air condenses and covers their bodies with small dewdrops. (Two photos composited into one). ▼

Ecological Aspects

Distribution of Dragonflies

Central European dragonflies are not evenly distributed. On the one hand, they fly in different areas and on the other, many are adapted to specific biotopes.

Vertical distribution

The damselflies such as Azure Bluets (*Coenagrion puella*) usually fly close to the water and between the plants, while dragonflies such as Blue Emperors (*Anax imperator*) tend to stay in the air above.

Stream and standing water dragonflies

Dragonflies can be divided into two groups according to their habitat requirements: the flowing water dragonflies and those that colonize standing water. Around a dozen species such as Hawkers or Featherlegs occur in both habitats. Rivers provide suitable habitats for running water dragonflies and the oxbow lakes separated from them for standing water species. According to counts from Dijkstra's identification book, 50 species live in standing water in Central Europe and only around 15 in flowing water. And this is also where many larvae burrow. Typical stream dragonflies are Banded Demoiselles (*Calopteryx splendens*) in the middle course, Beautiful Demoiselles (*Calopteryx virgo*) in the upper course, Goldenrings (*Cordulegaster*), or Clubtails (*Gomphidae*), whereby species as Small Pincertails (*Onychogomphus forcipatus*), also colonize lake or sand pit banks. In Southern Europe, however, several other species of stream dragonflies live in limited, separate areas, for example 6 particular species of Goldenrings. The dispersal dynamics of running water species are rather conservative, as they cannot spread so easily because rivers are often far apart. Typical still or standing water species are Darters (*Sympetrum*), Downy Emeralds (*Cordulia*), Black-tailed Skimmers (*Orthetrum cancellatum*), Blue Emperors (*Anax imperator*), Brighteyes (*Erythromma*) and Broad-bodied Chasers (*Libellula depressa*) or Four-spotted Chasers (*Libellula quadrimaculata*). Many Eurasian Bluetails (*Coenagrion*) also prefer still waters. The high number of species is due to the abundance of niches there. The dimensions of the water bodies, their vegetation, exposure to sun and wind, as well as the acidity of the water create a variety of conditions in stagnant waters to which very different dragonfly species are adapted. There are many very rare dragonfly species in Central Europe. These are adapted to special habitats such as bog lakes and ponds, which have been severely degraded. It may come as a surprise, but many dragonflies can be found in urban habitats.

Climate Fluctuations Can Change Dragonfly Behaviour

Hypothetical diagram of the reproduction of the Banded Demoiselle in two years from May (left) to September. In the upper year, the weather was consistently just warm enough for the damselflies to show threatening flights and territoriality as well as courtship (schematic drawings) at low and medium densities (height of the curve). Oviposition above the water surface was repeatedly observed. In the lower year it was cool for a long time and then very warm in July and August. Many Demoiselles emerged simultaneously from the larvae and

May   June   July   August   September

could only mate by fighting and chasing females. There were also many egg-laying events here, but now with many going underwater.

Dragonflies as Environmental Indicators

Dragonflies are good indicators of the condition of a body of water and its surroundings through their presence, especially if they are proving to breed there. Their larvae consume many filtering small invertebrate animals such as water fleas or particle eaters such as mayfly and stonefly larvae, animals that are important for biological water purification. If, for example, many Demoiselles fly over a body of water, it also means that their eggs and larvae are in the water as well as the many animals they feed on, without which turnover of waste substances and water purification would not work. Their occurrence as larvae in the water and in the air as flying adults provide information about the conditions of the micro habitats needed for both stages in their life cycle. Many dragonfly observers record the occurrence of dragonflies and this enables us document their spread and carry out research into the factors affecting their ongoing survival. Therefore, dragonflies also provide valuable information about climate-related shifts in populations.

Thus, more southern species such as the Scarlet Darter (*Crocothemis erythrea*) are appearing more and more in Central Europe. Also, continental species are migrating to Western Europe like the Green Snaketail (*Ophiogomphus cecilia*), as the climate is getting warmer. Cold-loving and northern species such

◀ The Scarlet Dragonfly (*Crocothemis erythrea*), which originates from the Mediterranean region, has already made its way to Northern Germany – here in a heat-reducing, so-called obelisk posture.

as some of the Striped Emeralds (*Somatochlora species*) such as the ice age relic, the Alpine Emerald (*Somatochlora alpestris*) are moving to higher altitudes in the mountains and to higher latitudes further north. Especially also running water dragonflies, which are adapted to a narrower temperature range than still water species, suffer from temperature increases. So, there are winners and losers among the dragonflies in climate change, as the group around Frank Suhling has worked out.

What We Can Do

No other order of insects is as comprehensively observable as that of the dragonflies. Beetles are hardly seen; butterflies fly far or dipterans are usually very small. Dragonflies, are large, conspicuous insects and often in the air in a confined space. On a few square meters, you can already see the entire behaviour of damselflies, while the action space of dragonflies expands, but they too always return to the water. Therefore, a very useful motto to plan intimate encounters with dragonflies is:

"Do not rush past, but linger"

This is unusual, because we humans are obviously runners. In Japan, our colleagues wanted to know how we film. So, six of them sat by our side. After an hour, only one was left, all the others had gone dragonfly hunting. On excursions, we also suggested sitting quietly at good spots for a long time. Many students subsequently reported interesting observations.

Binoculars with close focusing and cameras with medium-length telephoto or macro lenses are recommended. Even better would be to join a specialist organization which exist in many countries like the British Dragonfly Society (BDS), or Dragonfly Society of the Americas (DSA), or one of the many local groups, too. In such communities, larger projects such as the creation of a larger body of water can also be realised.

We would, however, like to recommend smaller garden ponds, too: almost 20 species of dragonflies and many other water-bound creatures have appeared at our 10 x 4m garden pond. About 25 % of the photos in this book were taken there and very many interesting observations were made.

List of Dragonfly Species Covered in the Book

Scientific Name	British and Irish Name	European Common Name
Filmed in Europe		
Chalcolestes viridis	Willow Emerald Damselfly	Western Willow Spreadwing
Lestes sponsa	Emerald Damselfly	Common Spreadwing
Calopteryx splendens	Banded Demoiselle	Banded Demoiselle
Calopteryx virgo	Beautiful Demoiselle	Beautiful Demoiselle
C. xanthostoma	Western Demoiselle	Western Demoiselle
C. haemorrhoidalis	Copper Demoiselle	Copper Demoiselle
Platycnemis latipes		White Featherleg
Platycnemis pennipes	White-legged Damselfly	Blue Featherleg
Pyrrhosoma nymphula	Large Red Damselfly	Large Red Damsel
Coenagrion puella	Azure Damselfly	Azure Bluet
Enallagma cyathigerum	Common Blue Damselfly	Common Bluet
Erythromma viridulum	Small Red-eyed Damselfly	Small Redeye
Ischnura elegans	Blue-tailed Damselfly	Common Bluetail
Aeshna cyanea	Southern Hawker	Blue Hawker
Aeshna mixta	Migrant Hawker	Migrant Hawker
Anax imperator	Emperor Dragonfly	Blue Emperor
Gomphus vulgatissimus	Common Clubtail	Common Clubtail
Cordulegaster boltonii	Golden-ringed Dragonfly	Common Goldenring
Cordulia aenea	Downy Emerald	Downy Emerald
Libellula depressa	Broad-bodied Chaser	Broad-bodied Chaser
Libellula fulva	Scarce Chaser	Blue Chaser
Libellula quadrimaculata	Four-spotted Chaser	Four-spotted Chaser
Orthetrum cancellatum	Black-tailed Skimmer	Black-tailed Skimmer
Sympetrum striolatum	Common Darter	Common Darter
Crocothemis erythraea	Scarlet Darter	Broad Scarlet
Filmed abroad		
Neurothemis fluctuans	Dark Red Bishop	
Neurobasis chinensis	Metalwing Demoiselle	
Tramea lacerata	Black Saddleback	
Anax junius	Green Darner	Common Green Darner
Rhyothemis fuliginosa	Butterfly Dragonfly	

Correction to: Larvae

Correction to:
Chapter 27 in: G. Rüppell, D. Hilfert-Rüppell, *Dragonfly Behavior,*
https://doi.org/10.1007/978-3-662-70234-5_27

The original version of the book was inadvertently published with the text appearing in German in the Figure on p.147 of Chapter 27. The figure has been replaced with the text appearing in English as shown below.

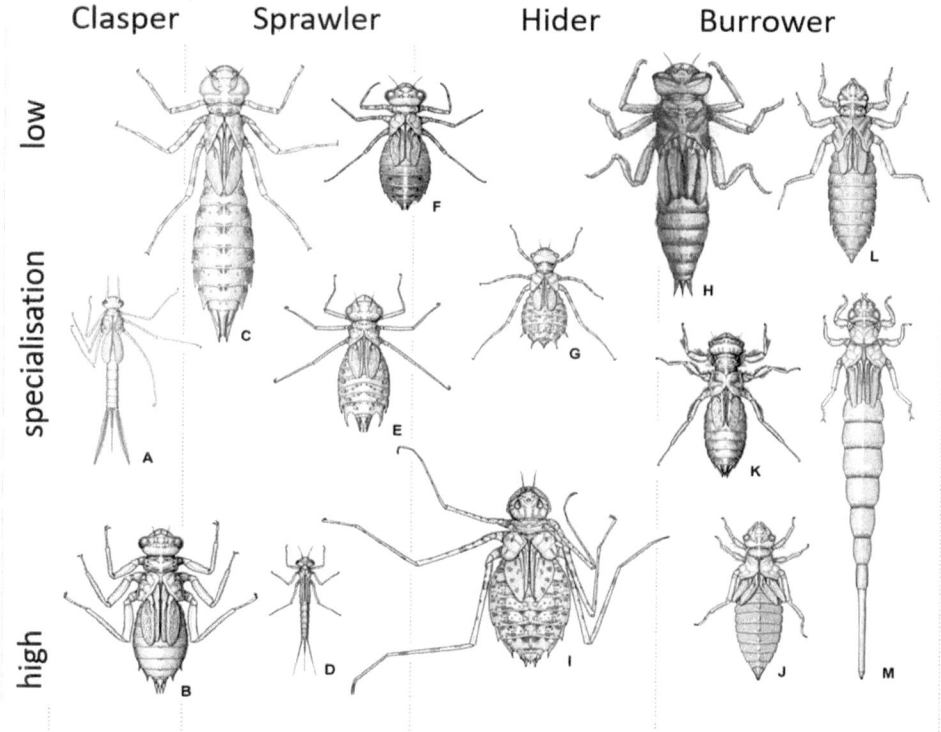

The correction chapter and the book have been updated.

The updated version of this chapter can be found at
https://doi.org/10.1007/978-3-662-70234-5_27

THANKS

We would like to thank our families, who often shook their heads at our long-term preoccupation with dragonflies. But when we were once invited to China to report on dragonfly flight and were royally entertained, our son Jan paused during the opulent meal and said: "Your dragonflies aren't that bad". As a six-year-old, he had already written: "Calopteryx, that doesn't mean anything to me!" (Calopteryx, das sagt mir nix!). This was symbolic for our personal environment at a time when dragonflies had become a passion for us. Dagmar's brother Michael, my daughter Maren and son Olav also came along and helped when it came to rescuing crashed dragonflies or to carry out recordings or investigations in Southern France.

We owe many stimulating conversations and assistance to the members of the Braunschweig dragonfly group at the Zoological Institute. Especially Frank Suhling, Andreas Martens and Gunnar Rehfeldt were the pillars in this wonderful team. But also, many students who wrote their final thesis with us, did not shy away from wet feet or hours of patient sitting at any waters in Northern Germany or Southern France. Frank extended his research areas to Namibia and Andreas was often on islands looking for dragonflies. Otto und Nils Larink, Olav and Maren Rüppell, Hansruedi Wildermuth, Klaus Borchert, Frank Suhling and Birgit Jarosch helped to improve the manuscript. Stephan Meyer made the layout and Stefanie Wolf, Meike Barth and Lars Koerner, for the English edition, gave the project a lot of propulsion. Lauren Kennedy from the BDS helped with the naming of the dragonflies. And luckily Steve Cham polished our English.

External dragonfly researchers have always accompanied our research and film recordings. Here I would like to mention the late Professor Peter Miller from Oxford, who classified our results or provided new aspects through encouraging and goal-oriented questions and discussions. The same applied to the late Professor Norman Moore from Swansey, whose work was more oriented towards nature conservation. I would like to name the late Professor Philip Corbet as a third intellectual father of ours, with whom we discussed for hours on frequent Scotland excursions. I remember an event at the University of Edinburgh

G. Rüppell, D. Hilfert-Rüppell, *Dragonfly Behavior*, https://doi.org/10.1007/978-3-662-70234-5

where he talked about the creation of his classic book on all aspects of dragonfly life to students for 2 hours. You could have heard a pin drop, so captivated and spellbound was the young audience. In Panama, Ola Fincke introduced me together with Gunnar Rehfeldt to the world of Giant Damselflies and patiently rejoiced over even the smallest progress for weeks. In Japan, it was Kiyoshi Inoue and Hidenori Ubukata who helped us together with Rainer Rudolph and Gunnar Rehfeldt to experience this diverse dragonfly world there. Rainer Rudolph was there in Japan and later also in Texas together with Heike Hadrys committed companion. Many members of the German Society for Odonatology (GDO) have enriched us again and again with their questions and suggestions at the annual meetings. Outstanding were those of Hansruedi Wildermuth or Reinhard Jödicke, as well as André Günther, who also accompanied us to Thailand.

Finally, we are also indebted to many representatives of authorities who allowed us to work in protected areas. This applies especially to the districts Gifhorn and Wolfenbüttel, as well as for the Environmental Office of the city of Braunschweig. We also had a very rich workplace in the stone steppe Crau in Southern France on the Canal de Vergière with over 40 species of dragonflies. At the local Ecomusée in St.Martin-de-Crau, we always had contacts and help. We were even able to organize a large-format dragonfly exhibition there and give a lecture in French, which, however, probably convinced more through images than through our bumpy language. Thanks also to the former Institute for Scientific Film in Göttingen and the popular television chain ZDF for enabling us to carry out our initially very costly slow-motion recordings in the non-digital era. Finally, Doclights and the television NDR helped us to show the beauty and importance of dragonflies to a wide audience in a film.

THE AUTHORS

Georg Rüppell and Dagmar Hilfert-Rüppell have lived and worked together for over 30 years. They have made three television films and two books about dragonflies. They both share a passion for researching, photographing and filming nature. Their shooting technique is characterized by a lot of patience and quick reactions with moving cameras. Their son Jan also brought this to the table as a drone filmmaker, who particularly enjoyed the difficult flight passages.

Photographic and Filmic Remarks

The images in this book come from various recording devices. Digital Canon cameras were used as photo cameras, mostly with medium-length macro lenses. Only the higher resolutions of the film cameras from Full HD allowed single images from the slow-motion films to be used. Dragonflies at rest are approached slowly. They are best depicted with medium focal length macro lenses (100–200 mm). It is advantageous if the insects sit at a right angle or perpendicular to the camera with the body's longitudinal axis, as then many body parts are sharply depicted. A small aperture helps to increase the depth of field.

For dragonflies sitting still, for example in the morning, many shots can be taken one after the other from a tripod with the camera on an adjustment slide with slightly shifted focus levels. With appropriate software, these images are combined into a very sharp single image (stacking). Flying dragonflies are best photographed without a tripod and with a moving camera in burst mode. To get them sharp, very short exposure times of less than 1/2000 sec are necessary. To get them into the picture, we use two methods.

We look at dragonflies on a constant flight path or hovering in place through the viewfinder and take many shots in a row in series. We try to capture unpredictable flight manoeuvres without looking through the viewfinder. Then we focus in advance on the probable shooting location and also take a series of shots, relying on good luck. By following the camera with the dragonfly movement, the number of waste shots can be significantly reduced.

Both methods also apply to filming, only then the camera runs the whole time. We start it even before the dragonflies have reached the focus levels. With all shots, we try to be as low as possible with the cameras, i.e. at eye level with the dragonflies, in order to get only slight distortions. Thus, working from a canoe has proven successful. The advantage of slow-motion film recordings is that all conceivable phases of behaviour can be recorded and used as single images. The shorter the exposure time, the sharper the individual images. Many images in this book come from such films.

LITERATURE

General

Abbott J C, Bota-Sierra C A, Guralnick R, Kalkman V, González-Soriano E, Novelo-Gutiérrez R ... Belitz M W (2022) Diversity of Nearctic dragonflies and damselflies (Odonata). Diversity 14(7):575

Alcock J (2009) Animal Behavior: An Evolutionary Approach /9. Sinauer Publishers

Askew R R (2021) The dragonflies of Europe. Brill

Boudot J P, Doucet G, Grand D (2021) Dragonflies and Damselflies of Britain and Western Europe: A Photographic Guide. Bloomsbury Publishing

Brockhaus T, Fischer U (2005) Die Libellenfauna Sachsens. Natur & Text, Rangsdorf, ISBN 978-3-9810058-0-6

Bybee S M, Kalkman V J, Erickson R J, Frandsen P B, Breinholt J W, Suvorov A ... Ware J L (2021) Phylogeny and classification of Odonata using targeted genomics. Molecular phylogenetics and evolution 160:107115

Carle F, Kjer K, May M (2015) A molecular phylogeny and classification of *Anisoptera* (Odonata). Arthropod Systematics & Phylogeny 73:281-301

Corbet P S (1999) Dragonflies: Behaviour and Ecology of Odonata. Harley Books, Colchester, ISBN 0-946589-64-X

Córdoba-Aguilar A, Cordero-Rivera A (2005) Evolution and ecology of Calopterygidae (Zygoptera: Odonata): status of knowledge and research perspectives. Neotropical entomology 34:861–879

Córdoba-Aguilar A, Beatty C, Bried J T (eds) (2023) Dragonflies and damselflies: model organisms for ecological and evolutionary research. Oxford University Press

Dijkstra K D, Kalkman V J (2012) Phylogeny, classification and taxonomy of European dragonflies and damselflies (Odonata): a review. Organisms Diversity & Evolution 12(3):209–227

Dijkstra K D, Schröter A (2020) Field guide to the dragonflies of Britain and Europe. Bloomsbury Publishing

Dumont H J, Vierstraete A, Vanfleteren J R (2010) A molecular phylogeny of the Odonata (Insecta). Systematic Entomology 35(1):6–18

Dreyer W (1986) Die Libellen. Gerstenberg, Hildesheim

Fabre J H (1924–1925) Souvenirs entomologiques. Paris Delagrave

Gnatzy W, Tautz J (2023) Insekten – Erfolgsmodelle der Evolution – faszinierend und bedroht. Springer, Berlin

Hilfert-Rüppell D, Rüppell G (2007) Juwelenschwingen / Gossamer Wings. Splendens, Cremlingen, ISBN 3-00-020389-3

Jödicke R (1997) Die Binsenjungfern und Winterlibellen Europas. Westarp Wissenschaften ISBN: 978-3-89432-460-5

Jurzitza G (2000) Der Kosmos-Libellenführer. Franckh-Kosmos, Stuttgart, ISBN 3-440-08402-7

Kappeler P M (2020) Verhaltensbiologie. 5. Springer, ISBN 978-3662605455

Karjalainen S, Hämäläinen M (2013) Neidonkorennot – Demoiselle Damselflies. Caloptera Publishing

Lehmann A, Nüß J H (2016) Libellen – Bestimmungsschlüssel für Nordeuropa, Mitteleuropa und Frankreich. Deutscher Jugendbund für Naturbeobachtung, Hamburg, ISBN 978-3-923376-15-5

Martens A (1996) Die Federlibellen Europas (Platycnemididae). Westarp & Spektrum, Magdeburg, Heidelberg, Berlin, Oxford, ISBN 3-89432-458-9

Martin G, Thévenon E (1994) Les libellules. La vie secrète des filles de l'air. Èditions de La Martinière, Paris

Mill P, Brooks S, Parr A (2010) Dragonflies (Odonata) in Britain and Ireland. In: Silent summer: The state of wildlife in Britain and Ireland, p. 471

Miller P L (1994) Dragonflies. The Richmond Publishers, Slough

Paulson D (2019) Dragonflies and Damselflies – a natural history. Princeton University Press

Robert P A (1959) Die Libellen (Odonaten). Kümmerly & Frey, Bern

Rüppell G, Hilfert-Rüppell D, Rehfeldt G, Schütte C (2005) Die Prachtlibellen Europas, Westarp Wissenschaften, Hohenwarsleben, ISBN 3-89432-883-5

Schiemenz H (1953) Die Libellen unserer Heimat. Urania, Berlin

Sternberg K, Buchwald R (1999) Die Libellen Baden-Württembergs 1, Allgemeiner Teil, Kleinlibellen (Zygoptera). Ulmer, Stuttgart

Sternberg K, Buchwald R (2000) Die Libellen Baden-Württembergs 2, Großlibellen (Anisoptera), Literatur. Ulmer, Stuttgart

Silsby J (2001) Dragonflies of the World. CSIRO publishing

Suhling F, Müller O (1997) Die Flußjungfern Europas, Westarp Wissenschaften, Hohenwarsleben, ISSN 0940-6638

Suhling F, Martens A (2007) Dragonflies and damselflies of Namibia. Windhoek: Gamsberg Macmillan

Suhling F, Sahlén G, Gorb S, Kalkman V J, Dijkstra K D B, van Tol J (2015) Order odonata. In: Thorp J H (ed) Thorp and Covich's freshwater invertebrates. Academic Press, pp 893–932

Svensson E I, Kristoffersen L, Oskarsson, K & Bensch S (2004) Molecular population divergence and sexual selection on morphology in the banded demoiselle (*Calopteryx splendens*). *Heredity*, *93*(5), 423–433

Svensson E I, Karlsson K, Friberg M & Eroukhmanoff F (2007) Gender differences in species recognition and the evolution of asymmetric sexual isolation. Current Biology, 17(22), 1943–1947

Wildermuth H, Martens A (2019) Die Libellen Europas. Quelle & Meyer, Wiebelsheim, ISBN 978-3-494-01690-0

Wilson K (2009) Dragonfly giants. Agrion 13:29–31

Wings

Appel E, Michels J, Gorb S N (2023) Resilin in Insect Flight Systems. Advanced Functional Materials, 2215162

Gorb S N (1999) Serial elastic elements in the damselfly wing: mobile vein joints contain resilin. Naturwissenschaften 86:552–555

Johansson F, Söderquist M, Bokma F (2009) Insect wing shape evolution: independent effects of migratory and mate guarding flight on dragonfly wings. Biological Journal of the Linnean Society 97(2):362–372

Jongerius S R, Lentink D (2010) Structural analysis of a dragonfly wing. Experimental Mechanics 50_1323-1334

Pfau H K (1986) Untersuchungen zur Konstruktion, Funktion und Evolution des Flugapparates der Libellen (Insecta, Odonata). Tijdschrift voor entomologie 129(3):35–123

Rajabi H, Gorb S N (2020) How do dragonfly wings work? A brief guide to functional roles of wing structural components. International Journal of Odonatology 23(1):23–30

Ren L, Li X (2013) Functional characteristics of dragonfly wings and its bionic investigation progress. Science China Technological Sciences 56:884–897

Wootton R J (1992) Functional morphology of insect wings. Annual review of entomology 37(1):113–140

Wootton R J, Newman D J (2008) Evolution, diversification, and mechanics of dragonfly wings. Dragonflies & damselflies. Model organisms for ecological and evolutionary research 261:274

Flight

Azuma A & Watanabe T (1988) Flight performance of a dragonfly. Journal of experimental biology, 137(1), 221-252

Bode-Oke A T, Zeyghami S, Dong H (2018) Flying in reverse: kinematics and aerodynamics of a dragonfly in backward free flight. Journal of The Royal Society Interface 15(143): 20180102

Bomphrey R J, Nakata T, Henningsson P, Lin H T (2016) Flight of the dragonflies and damselflies. Philosophical Transactions of the Royal Society B: Biological sciences 371(1704): 20150389

Dong H, Liang Z, Wan H, Koehler C, Gaston Z (2010, June) An integrated analysis of a dragonfly in free flight. In: 28th AIAA applied aerodynamics conference, p 4390

Ellington C P (1999) The novel aerodynamics of insect flight: applications to micro-air vehicles. Journal of Experimental Biology 202:3439–3448

Grabow K, Rüppell G (1995) Wing loading in relation to size and flight characteristics of European Odonata. Odonatologica 24(2):175–186

Hilfert-Rüppell D, Rüppell G (2013) Do coloured-winged damselflies and dragonflies have flight kinematics different from those with clear wings? International Journal of Odonatology 16(2):119–134, DOI:10.1080/13887890.2013.763332

Hilfert-Rüppell D (2015) High frequency and counterstroking: *Calopteryx splendens* female threatening flight. International Journal of Odonatology 18(1):55–64, DOI:10.1080/13887 890.2015.1013511

Hu Z & Deng X Y (2014) Aerodynamic interaction between forewing and hindwing of a hovering dragonfly. Acta Mechanica Sinica, 30(6), 787-799

Huang H & Sun M (2007) Dragonfly forewing-hindwing interaction at various flight speeds and wing phasing. AIAA journal, 45(2), 508-511

Kesel A B (1998) Biologisches Vorbild Insektenflügel – Mehrkriterienoptimierung ultraleichter Tragflächen. BIONA-report 12:107–111

Koehler C, Liang Z, Gaston Z, Wan H, Dong H (2012) 3D reconstruction and analysis of wing deformation in free-flying dragonflies. Journal of Experimental Biology 215(17):3018–3027

Lehmann F O (2009) Wing–wake interaction reduces power consumption in insect tandem wings. Experiments in fluids 46:765–775

Li C, Dong H (2017) Wing kinematics measurement and aerodynamics of a dragonfly in turning flight. Bioinspiration & biomimetics 12(2):026001

May M L (1995) Dependence of flight behavior and heat production on air temperature in the green darner dragonfly *Anax junius* (Odonata: Aeshnidae). Journal of Experimental Biology 198(11):2385–2392

Nachtigall W (1977) Zur Bedeutung der Reynoldszahl und der damit zusammenhängenden strömungsmechanischen Phänomene in der Schwimmphysiologie und Flugbiophysik. Fortschr. Zool. 24:14–56

Nachtigall W, Wisser A (2003) Insektenflug. Springer, Berlin

Norberg R Å (1972) The pterostigma of insect wings an inertial regulator of wing pitch. J Comp Physiol 81:9–22

Norberg U L (2007) Evolution of flight in animals. Flow Phenom Nat 1:36–48

Rudolph R (1976) Some aspects of wing kinematics in *Calopteryx splendens* (Harris) (Zygoptera: Calopterygidae). Odonatologica 5(2):119–127

Rudolph R (1977) Aerodynamic properties of *Libellula quadrimaculata* L. (Anisoptera: Libelludidea), and the flow around smooth and corrugated wing section models during gliding flight. Odonatologica 7:49–58

Rüppell G (1982) Vogelflug. Rowohlt Taschenbuch, Reinbeck bei Hamburg

Rüppell G (1985) Kinematic and behavioural aspects of flight of the male Banded Agrion, *Calopteryx* (Agrion) *splendens* L. In: Gewecke L M, Wendler G (eds) Insect Locomotion. Parey, Berlin, p 195–204

Rüppell G (1989) Kinematic analysis of symmetrical flight manoeuvres of Odonata. Journal of Experimental Biology 144:13–42

Rüppell G (1992) Über den Flug und die Fortpflanzung der Libellen. Biologie in unserer Zeit, 22(4):196–202

Rüppell G, Hilfert D (1993) The flight of the relict dragonfly *Epiophlebia superstes* (Selys) in comparison with that of the modern Odonata (Anisozygoptera: Epiophlebiidae). Odonatologica 22(3):295–309

Rüppell G, Hilfert-Rüppell D (2010) Kinematic analysis of maiden flight of Odonata. International Journal of Odonatology 13(2):181–192, DOI:10.1080/13887890.2010.9748373

Rüppell G, Hilfert-Rüppell D (2020) Rapid acceleration in Odonata flight: highly inclined and in-phase wing beating. International Journal of Odonatology 23(1):63–78, DOI:10.1080/13 887890.2019.1688017

Rüppell G, Hilfert-Rüppell D (2023) Double function of flight in *Calopteryx splendens* males. International Journal of Odonatology 26:172–179, DOI:10.48156/1388.2023.1917232

Salami E, Ward T A, Montazer E, Ghazali N N N (2019) A review of aerodynamic studies on dragonfly flight. Proceedings of the Institution of Mechanical Engineers, Part C. Journal of Mechanical Engineering Science 233(18):6519–6537

Samejima Y, Tsubaki Y (2010) Body temperature and body size affect flight performance in a damselfly. Behavioral Ecology and Sociobiology 64:685–692

Sun M & Lan S L (2004) A computational study of the aerodynamic forces and power require-ments of dragonfly (*Aeschna juncea*) hovering. Journal of Experimental Biology, 207(11), 1887-1901

Sun X, Gong X & Huang D (2017) A review on studies of the aerodynamics of different types of maneuvers in dragonflies. Archive of Applied Mechanics, 87, 521-554

Usherwood J R & Lehmann F O (2008) Phasing of dragonfly wings can improve aerodynamic efficiency by removing swirl. Journal of The Royal Society Interface, 5(28), 1303-1307

Wang H, Zeng L, Liu, H & Yin C (2003) Measuring wing kinematics, flight trajectory and body attitude during forward flight and turning maneuvers in dragonflies. Journal of Experimen-tal Biology, 206(4), 745-757

Wang Z J (2004. The role of drag in insect hovering. Journal of Experimental Biology, 207(23), 4147-4155

Wang Z J (2008) Dragonfly flight. Physics today 61(10):74–75

Weis-Fogh T (1973) Quick estimates of flight fitness in hovering animals, including novel mechanisms for lift production. Journal of experimental Biology 59(1):169–230

Wootton R J, Kukalová-Peck J (2000) Flight adaptations in Palaeozoic Palaeoptera (Insecta). Biological Reviews 75(1):129–167

Wootton R J (2020) Dragonfly flight: morphology, performance and behaviour. International Journal of Odonatology 23(1):31–39

Prey Capture

Combes S A, Rundle D E, Iwasaki J M, Cral, J D (2012) Linking biomechanics and ecology through predator-prey interactions: flight performance of dragonflies and their prey. Jour-nal of Experimental Biology 215(6):903–913

Combes S A, Salcedo M K, Pandit M M, Iwasak, J M (2013) Capture success and efficiency of dragonflies pursuing different types of prey. Integrative and comparative biology 53(5):787–798

Kundanati L, Das P, Pugno N M (2019) Prey capturing and feeding apparatus of dragonfly nymph. bioRxiv, 536805

Olberg R M, Worthington A H, Venator K R (2000) Prey pursuit and interception in dragonflies. Journal of Comparative Physiology A 186:155–162

Olberg R M, Worthington A H, Fox J L, Bessette C E, Loosemore M P (2005) Prey size selection and distance estimation in foraging adult dragonflies. Journal of Comparative Physiology A, 191:791–797

Olberg R M, Seaman R C, Coats M I, Henry A F (2007) Eye movements and target fixation during dragonfly prey-interception flights. Journal of comparative physiology A 193:685–693

Pritchard G (1965) Prey capture by dragonfly larvae (Odonata; Anisoptera). Canadian Journal of Zoology 43(2):271–289

Rüppell G (1999) Prey capture flight of *Calopteryx haemorrhoidalis* (Van der Linden) (Zygoptera: Calopterygidae). International Journal of Odonatology 2(1):123–131

Fighting

Baird J M, May M L (2003) Fights at the dinner table: agonistic behavior in *Pachydiplax longipennis* (Odonata: Libellulidae) at feeding sites. Journal of Insect Behavior 16:189–216

Dos Santos T B, Peixoto P E C (2017) Agonistic interactions in the dragonfly *Micrathyria ungulata*: does male fighting investment come from an innate ability or an indomitable will? Behavioral Ecology and Sociobiology 71:1–9

Kasuya E, Edanami K, Ohno I (1997) Territorial conflicts in males of the dragonfly, *Orthetrum japonicum japonicum* (Odonata: Libellulidae): the role of body size. Zoological science 14(3):505–509

Moore N W (2000) Interspecific encounters between male aeshnids do they have a function? International Journal of Odonatology 3(2):141–151

Pajunen V (1966) Aggressive behaviour and territoriality in a population of *Calopteryx virgo* L. (Odonata: Calopterygidae). Annales Zoolica Fennica 3:201–214

Rillich J, Stevenson P A (2019) Fight or flee? Lessons from insects on aggression. Neuroforum 25(1):3–13

Rüppell G, Hilfert-Rüppell D (2013) Biting in dragonfly fights. International Journal of Odonatology 16(3):219–229, DOI:10.1080/13887890.2013.804364

Rüppell G (1989) Fore legs of dragonflies used to repel males. Odonatologica 18(4):391–396

Singer F (1989) Interspecific aggression in *Leucorrhinia* dragonflies: a frequency-dependent discrimination threshold hypothesis. Behavioral ecology and sociobiology 25:421–427

Switzer P V (2004) Fighting behavior and prior residency advantage in the territorial dragonfly, *Perithemis tenera*. Ethology Ecology & Evolution 16(1):71–89

Reproduction

Anders U, Rüppell G (1997) Zeitanalyse der Balzflüge europäischer Prachtlibellen-Arten zur Betrachtung ihrer Verwandtschaftsbeziehungen (Odonata: Calopterygidae). Entomologia Generalis 21(4):253–264

Andersson M. (1994) Sexual Selection. Princeton University Press, Princeton

Cham S (2012) A study of the Southern Hawker *Aeshna cyanea* emergence for a garden pond. Journal of the British Dragonfly Society, 28(1), 1–20

201

Clutton-Brock TM (1988) Reproductive Success. University of Chicago Press, Chicago, London

Conrad KF, Pritchard G (1992) An ecological classification of odonate mating systems: The relative influence of natural, inter- and intra-sexual selection. Biological Journal of the Linnean Society 45:255–269

Córdoba-Aguilar A, González-Tokman D, Nava-Bolaños D Á, Cuevas-Yáñez K, Rivas M, Nava-Sánchez A (2015) Female choice in damselflies and dragonflies. Cryptic Female Choice in Arthropods: Patterns, Mechanisms and Prospects 239–253

Fincke O M (1982) Lifetime mating success in a natural population of the damselfly, *Enallagma hageni* (Walsh) (Odonata: Coenagrionidae). Behavioral Ecology and Sociobiology 10:293–302

Fincke O M (1984) Giant damselflies in a tropical forest: reproductive behavior of *Megaloprepus coerulatus* with notes on *Mecistogaster*. Advances in Odonatology 2:13–27

Fincke O M (1992) Interspecific competition for tree holes: consequences for mating systems and coexistence in neotropical damselflies. American Naturalist 139:80–101

Fincke O M (1997) Conflict resolution in the Odonata: implications for understanding female mating patterns and female choice. Biological Journal of the Linnean Society 60(2):201–220

Fincke O M, Waage J K, Koenig W D (1997) Natural and sexual selection components of odonate mating patterns. In: Choe J C, Crespi BJ (eds) The evolution of mating systems in insects and arachnids. Cambridge University Press

Günther A (2015) Signalling with clear wings during territorial behaviour and courtship of *Chlorocypha cancellata* (Odonata, Chlorocyphidae). International Journal of Odonatology 18(1):45–54

Günther A (2008) Vergleichende Untersuchungen zum Reproduktionsverhalten südostasiatischer Chlorocyphidae und Calopterygidae (Odonata: Zygoptera). Ph-thesis, TU Freiberg

Günther A (2019) Reproductive behaviour of Chlorocyphidae. Part 1. Genus *Sclerocypha* Fraser, 1949 (Odonata). Odonatologica 48(3–4):285–304

Günther A (2021) Reproductive behaviour of Chlorocyphidae. Part 2. Genus *Disparocypha* Ris, 1916 (Odonata). Odonatologica 49(1–2):85–106

Günther A, Hilfert-Rüppell D, Rüppell G (2014) Reproductive behaviour and the system of signalling in *Neurobasis chinensis* (Odonata, Calopterygidae) – a kinematic analysis. International Journal of Odonatology 17(1):31–52

Hilfert D, Rüppell G (1997) Alternative mating tactics in *Calopteryx splendens* (Odonata: Calopterygidae). Advances in Ethology 32:47

Hilfert D (1997) Motivation as a mechanism to optimise mating success in *Calopteryx* (Odonata: Calopterygidae). Advances in Ethology 32:235

Hilfert D, Rüppell G (1997) Alternative mating tactics in *Calopteryx splendens* (Odonata: Calopterygidae). Mitteilungen der Deutschen Gesellschaft für allgemeine und angewandte Entomologie 11:411–414

Hilfert-Rüppell D (2000) To stay or not to stay: decision-making during territorial behaviour of *Calopteryx haemorrhoidalis* and *Calopteryx splendens splendens* (Zygoptera: Calopterygidae). International Journal of Odonatology 2:167–175

Hilfert-Rüppell D (2004) Optimierung des Fortpflanzungsverhaltens: wichtige Einflußgrößen auf Territorialität und auf Paarungen von europäischen Prachtlibellenmännchen (Odonata: Zygoptera). Diss. Braunschweig, Techn. Universität

Hilfert-Rüppell D, Rüppell G (2009) Males do not catch up with females in pursuing flight in *Calopteryx splendens* (Odonata: Calopterygidae). International Journal of Odonatology 12(2):195–203

Kaiser H (1974) Verhaltensgefüge und Temporialverhalten der Libelle *Aeschna cyanea* (Odonata). Zeitschrift für Tierpsychologie 34(4):398–429

Kaunisto K M, Suhonen J (2023) Territorial males have larger wing spots than non-territorial males in the damselfly *Calopteryx splendens* (Zygoptera: Calopterygidae). International Journal of Odonatology 26:1–6

Martens A (2015) Alternative oviposition tactics in *Zygonyx torridus* (Kriby) (Odonata: Libellulidae): modes and sequential flexibility. International Journal of Odonatology 18:71–80

Michiels N K, Dhondt A A (1990) Costs and benefits associated with oviposition site selection in the dragonfly *Sympetrum danae* (Odonata: Libellulidae). Animal Behaviour 40(4):668–678 https://doi.org/10.1016/S0003-3472(05)80696-7

Miller A K, Miller P L, Siva-Jothy M T (1984) Pre-copulatory guarding and other aspects of reproductive behaviour in *Sympetrum depressiusculum* (Selys) at rice fields in southern France (Anisoptera: Libellulidae). Odonatologica 13(3):407–414

Parker GA (1970) Sperm competition and its evolutionary consequences in the insects. Biological Reviews 45:525–567

Poethke H J, Kaiser H (1987) The territoriality threshold: a model for mutual avoidance in dragonfly mating systems. Behavioral Ecology and Sociobiology 20:11–19

Plaistow S J, Siva-Jothy M T (1996) Energetic constraints and male mate securing tactics in the damselfly *Calopteryx splendens xantosthoma* (Charpentier). Proceedings of the Royal Society of London B 263:1233–1238

Plaistow S J, Tsubaki Y (2000) A selective trade-off for territoriality and nonterritoriality in the polymorphic damselfly *Mnais costalis*. Proceedings of the Royal Society of London B 267:969–975

Rüppell G, Hilfert-Rüppell D (2019) Touching water by males of *Calopteryx virgo* L. (Insecta: Odonata) in threatening display. International Journal of Odonatology 22(1):31–36, DOI: 10.1080/13887890.2018.1563917

Rüppell G, Hilfert-Rüppell D (2014) Slow-motion analysis of female refusal behaviour in drag-onflies. International Journal of Odonatology 17(4):199–215, DOI:10.1080/13887890.2014.972893

Samejima Y, Tsubaki Y (2010) Body temperature and body size affect flight performance in a damselfly. Behavioral Ecology and Sociobiology 64:685-692

Siva-Jothy M T, Tsubaki Y (1994) Sperm competition and sperm precedence in the dragonfly *Nanophya pygmaea*. Physiological Entomology 19(4):363–366

Siva-Jothy M T, Hooper R E (1995) The disposition and genetic diversity of stored sperm in females of the damselfly *Calopteryx splendens xanthostoma* (Charpentier). Proceedings of the Royal Society of London. Series B: Biological Sciences 259(1356):313–318

Verzijden M N, Scobell S K & Svensson E I (2014) The effects of experience on the develop-ment of sexual behaviour of males and females of the banded demoiselle (*Calopteryx splendens*). Behavioural processes, 109, 180–189

Waage J K (1979) Dual function of the damselfly penis: sperm removal and transfer. Science, 203(4383):916–918

Larvae

Cham S (2021) Egg hatching, prolarvae and larval development time of Chalcolestes viridis (Vander Linden) (Willow Emerald Damselfly) in Britain. Journal of the British Dragonfly Society, 37, 40-59

Corbet P, Suhling F, Soendgerath D (2006) Voltinism of Odonata: a review. International Jour-nal of Odonatology 9(1),:1–44, DOI:10.1080/13887890.2006.9748261

Jakob C, Suhling F (1999) Risky times? Mortality during emergence in two species of dragon-flies (Odonata: Gomphidae, Libellulidae). Aquatic insects 21(1):1–10

Johansson F, Suhling F (2004) Behaviour and growth of dragonfly larvae along a permanent to temporary water habitat gradient. Ecological Entomology 29(2):196–202

Johansson F, Mikolajewski D J (2008) Evolution of morphological defences. In Córdoba-Aguilar A et al (Eds.) (2023) *Dragonflies and damselflies: model organisms for ecological and evolu-tionary research*. Oxford University Press

Johnson D M (1991) Behavioral ecology of larval dragonflies and damselflies. Trends in Ecol-ogy & Evolution 6(1):8–13

Krishnaraj R, Pritchard G (1995) The influence of larval size, temperature, and components of the functional response to prey density on growth rates of the dragonflies *Lestes disjunctus* and *Coenagrion resolutum* (Insecta: Odonata). Canadian Journal of Zoology 73(9):1672–1680

Martens A, Kohl S, Wildermuth H (2023) Are anal spines of anisopteran larvae an antipredator device? A case study in *Boyeria irene* (Odonata: Aeshnidae). Odonatologica 52(1–2):49–60

Mikolajewski D J, Johansson F (2004) Morphological and behavioral defenses in dragonfly larvae: trait compensation and cospecialization. Behavioral Ecology 15(4):614–620

Mill P J, Pickard R S (1975) Jet-propulsion in anisopteran dragonfly larvae. Journal of Comparative Physiology 97(4):329–338

Norling U (1984) Photoperiodic control of larval development in *Leucorrhinia dubia* (Van der Linden): a comparison between populations from northern and southern Sweden (Anisoptera: Libellulidae). Odonatologica 13(4):529–550

Suhling F (1995) Temporal patterns of emergence of the riverine dragonfly *Onychogomphus uncatus* (Odonata: Gomphidae). Hydrobiologia 302:113–118

Suhling F (2001) Intraguild predation, activity patterns, growth and longitudinal distribution in running water odonate larvae. Archiv Hydrobiologie 151(1):1–15

Suhling F, Schenk K, Padeffke T, Martens A (2004) A field study of larval development in a dragonfly assemblage in African desert ponds (Odonata). Hydrobiologia 528:75–85

Suhling F, Sahlén G, Kasperski J, Gaedecke D (2005) Behavioural and life history traits in temporary and perennial waters: comparisons among three pairs of sibling dragonfly species. Oikos 108(3):609–617

Suhling F, Suhling I, Richter O (2015) Temperature response of growth of larval dragonflies – an overview. International Journal of Odonatology 18(1):15–30

Van Buskirk J (1989) Density-dependent cannibalism in larval dragonflies. Ecology 70(5):1442–1449

Predation

Johnson D M, Martin T H, Mahato M, Crowder L B, Crowley P H (1995) Predation, density dependence, and life histories of dragonflies: a field experiment in a freshwater community. Journal of the North American Benthological Society 14(4):547–562

Rüppell G, Hilfert-Rüppell D, Schneider B, Dedenbach H (2020) On the firing line –interactions between hunting frogs and Odonata. International Journal of Odonatology 23(3):199–217, DOI:10.1080/13887890.2020.1733328

Rehfeldt G E (1990) Anti-predator strategies in oviposition site selection of *Pyrrhosoma nymphula* (Zygoptera: Odonata). Oecologia 85:233–237

Rehfeldt G E (1992) Impact of predation by spiders on a territorial damselfly (Odonata: Calopterygidae). Oecologia 89(4):550–556

Rehfeldt G E (1991) The upright male position during oviposition as an antipredator response in *Coenagrion puella* (L.) (Zygoptera: Coenagrionidae). Odonatologica 20:69–74

Rolff J, Vogel C, Poethke H J (2001) Co-evolution between ectoparasites and their insect hosts: a simulation study of a damselfly–water mite interaction. Ecological Entomology 26(6):638–645

Stoks R, McPeek M A, Mitchell, J L (2003) Evolution of prey behavior in response to changes in predation regime: damselflies in fish and dragonfly lakes. Evolution 57(3):574–585

Svensson E I & Friberg M (2007) Selective predation on wing morphology in sympatric damselflies. The American Naturalist, 170(1), 101–112

Wildermuth H (2000) Totstellreflex bei Großlibellenlarven (Odonata). Libellula 19:17–39

Wohlfahrt B, Mikolajewski D J, Joop G, Suhling F (2006) Are behavioural traits in prey sensitive to the risk imposed by predatory fish? Freshwater Biology 51(1):76–84

Migration

Dumont H J, Hinnekint B O N (1973) Mass migration in dragonflies, especially in *Libellula quadrimaculata* L.: a review, a new ecological approach and a new hypothesis. Odonatologica 2(1):1–20

Feng H Q, Wu K M, Ni Y X, Ceng D F, Guo Y Y (2006) Nocturnal migration of dragonflies over the Bohai Sea in northern China. Ecological Entomology 31(5):511–520

Kharitonov A Y, Popova ON (2011) Migrations of dragonflies (Odonata) in the south of the West Siberian plain. Entomological review 91:411–419

Knoblauch A, Thoma M, Menz M H (2021) Autumn southward migration of dragonflies along the Baltic coast and the influence of weather on flight behaviour. Animal Behaviour 176:99–109

May M L, Matthews J H (2008) Migration in Odonata: a case study of *Anax junius*. In Córdoba-Aguilar et al (Eds.) (2023) Dragonflies and damselflies: model organisms for ecological and evolutionary research. Oxford University Press

May M L (2013) A critical overview of progress in studies of migration of dragonflies (Odonata: Anisoptera), with emphasis on North America. Journal of Insect Conservation 17:1–15

Wikelski M, Moskowitz D, Adelman J S, Cochran J, Wilcove D S, May M L (2006) Simple rules guide dragonfly migration. Biology letters 2:325–329

Ecological Aspects

Bowler D E, Eichenberg D, Conze K J, Suhling F, Baumann K, Benken T … Bonn A (2021) Winners and losers over 35 years of dragonfly and damselfly distributional change in Germany. Diversity and Distributions 27(8):1353–1366

Bried J, Ries L, Smith B, Patten M, Abbot J, Ball-Damerow J … White E (2020) Towards global volunteer monitoring of odonate abundance. BioScience 70(10):914–923

Cardoso P, Barton P S, Birkhofer K, Chichorro F, Deacon C, Fartmann T … Samways M J (2020) Scientists' warning to humanity on insect extinctions. Biological conservation 242:108426

Chovanec A, Raab R (1997) Dragonflies (Insecta, Odonata) and the ecological status of newly created wetlands – examples for long-term bioindication programms. Limnologica 27(3): 381–392

Clausnitzer V, Kalkman V J, Ram M, Collen B, Baillie J E, Bedjanič M ... Wilson K (2009) Odonata enter the biodiversity crisis debate: the first global assessment of an insect group. Biological conservation 142(8):1864–1869

Clausnitzer V, Dijkstra K D B, Koch R, Boudot J P, Darwall W R, Kipping J ... Suhling F (2012) Focus on African freshwaters: hotspots of dragonfly diversity and conservation concern. Frontiers in Ecology and the Environment 10(3):129–134

Goertzen D, Suhling F (2013) Promoting dragonfly diversity in cities: major determinants and implications for urban pond design. Journal of Insect Conservation 17:399–409

Flenner I D A, Richter O, Suhling F (2010) Rising temperature and development in dragonfly populations at different latitudes. Freshwater Biology 55(2):397–410

Hilfert D, Rüppell G (1997) Early morning oviposition of dragonflies with low temperatures for male-avoidance (Odonata: Aeshnidae, Libellulidae). Entomologia generalis 177–188

Kalkman V J, Boudot J P, Bernard R, De Knijf G, Suhling F, Termaat T (2018) Diversity and conservation of European dragonflies and damselflies (Odonata). Hydrobiologia 811(1):269–282

Kalkman V J, Boudot J P, Futahash R, Abbott J C, Bota-Sierra C A, Guralnick R ... Belitz M W (2022) Diversity of Palaearctic dragonflies and damselflies (Odonata). Diversity, 14(11): 966

Moore N W (1991) The development of dragonfly communities and the consequences of territorial behaviour: a 27year study on small ponds at Woodwalton Fen, Cambridgeshire, United Kingdom. Odonatologica 20(2):203–231

Ott J (2010) Dragonflies and climatic change – recent trends in Germany and Europe. BioRisk 5:253–286

Rehfeldt G E (1995) Natürliche Feinde, Parasiten und Fortpflanzung von Libellen. Aqua & Terra, Wolfenbüttel

Rolff J (1999) Parasitism increases offspring size in a damselfly: experimental evidence for parasite-mediated maternal effects. Animal Behaviour 58:1105–1108

Sahlén G, Bernard R, Rivera A C, Ketelaar R, Suhling F (2004) Critical species of Odonata in Europe. International Journal of Odonatology 7(2):385–398

Stoks R, McPeek M A (2003) Predators and life histories shape *Lestes* damselfly assemblages along a freshwater habitat gradient. Ecology 84(6):1576–1587

Svensson E I, Gómez-Llano M A, Torres A R & Bensch H M (2018) Frequency dependence and ecological drift shape coexistence of species with similar niches. The American Naturalist, 191(6), 691–703

Taylor P, Smallshire D, Parr A, Brooks S, Cham S, Colver E, Roy D (2021) State of Dragonflies 2021. British Dragonfly Society, UK

Wellenreuther M, Larson K W & Svensson E I (2012) Climatic niche divergence or conservatism? Environmental niches and range limits in ecologically similar damselflies. Ecology, 93(6), 1353–1366

WEBLINKS

British Dragonfly Society: https://british-dragonflies.org.uk

Dutch Dragonfly Association https://www.brachytron.nl

Finnish Dragonfly Society: https://korento.net

Gesellschaft deutschsprachiger Odonatologen: https://www.libellula.org

Iberian Group of Odonatology: http://gio.sea-socios.com/.a

International Journal of Odonatology, open access: https://worlddragonfly.org/ijo/

Notulae odonatologicae:
https://bioone.org/journals/notulae-odonatologicae/editorial-office

Odonatologica: https://www.odonatologica.com

Société Française d'Odonatologie:
https://libellules.org/sfo/societe-francaise-odonatologie.html

Società Italiana per lo Studio e la Conservazione delle Libellule Onlus:
https://www.odonata.it

Swedish Dragonfly Society: https://www.trollslandeforeningen.se/in-english/

Worldwide Dragonfly Association: https://worlddragonfly.org

Wazki Odonata Polski: https://wazki.pl

IMAGE CREDITS

The spiracle, opening of the tracheal system by Steve Valley   [P. 5],

the nano image of the surface of a dragonfly wing
by Elena Ivanova   [P. 12],

the flow simulation of the dragonfly was created by Habido Dong
from the University of Virginia   [P. 21],

the photo of *Aristocypha fenestrella* by André Günther   [P. 78],

the photos of the Common Bluets and the one next to it of Blue
Emperor by Beat Schneider (Beat Schneider animal films)   [P. 82, 83, P. 170],

the schematic representation of the interior of Demoiselles
by Michael Gradias   [P. 107],

the SEM image of the abdomen of a female dragonfly
by Wolfgang Dreyer   [P. 134],

the photo of *Vorticella* by Otto Larink   [P. 142],

the pictures of the Hobbies by Thomas Plack (Vogelarchiv)   [P. 164],

the images of the swarm of Four-spotted Chasers
by Maxim Koshkin   [P. 182, 183],

all aerial photographs by Jan Rüppell,

all other drawings and photos by the authors.

Some of the drawings made by the authors have already been published in the "International Journal of Odonatology", others in "Die neue Brehm-Bücherei – Die Prachtlibellen Europas", Westarp Wissenschaften.

ANECDOTES

FROM ORNITHOLOGIST TO ODONATOLOGIST

Curiosity seems to be a significant research motivator. At least it was for me. I always wanted to see and describe new things in nature. Initially, it was flying birds that had captivated me. But that changed.

My students turned me from birds to dragonflies, and this is how it happened: We were on one of the many excursions at the Aller, a river in the heath of Northern Germany, westwards of the little town Gifhorn. Heiko, an elder student, held a rope in his hand. We, twenty people, wanted to cross the river with an inflatable boat. He had attached this rope to the boat and somehow secured it on the other side. In any case, he could pull the boat over the river via a loop. Someone had to start and be the first to cross. That was supposed to be me. So, I climbed into the inflatable boat – but probably a bit too briskly, because the boat was pushed away from me and I could just hold onto the rope with both hands. Most pairs of eyes looked quite amused. I hung like a captured frog on the rope. Only my feet were in the boat, the body hung on the rope close above the water. Did Heiko sense a possible punch line? His forehead was furrowed, but his mouth was laughing. Should he loosen the rope and thus dunk me into the water? No, luckily, he kept the rope taut and pulled me back into a stable position. I appreciated that very much. Not because of this, he later became a professor of fish ecology in Bremen, no, he had simply become an expert. On the other side, I received the others, and we discussed the program as if nothing had happened.

Blue insects were flying all around us, in swarms. Then a student asked me: "What are those blue things there?" "They must be dragonflies!" I said, the bird expert. "I've heard something about Demoiselles" and then, he obviously knew my bird flight films, came the question of all questions: "Can you also film them in slow motion?" What? How? Don't you get it? Think! Yes, why not. That's when it happened: A new window opened – a window into a completely new world, into the world of dragonflies. Demoiselles have blue wings and for the first time their full beauty and the sequence of their movements were revealed through slow-motion recordings. At the same time, other members of my team had taken

G. Rüppell, D. Hilfert-Rüppell, *Dragonfly Behavior*, https://doi.org/10.1007/978-3-662-70234-5

dragonflies into their research focus: Gunnar, Andreas and Frank. We subsequently became a proper dragonfly working group, exchanging ideas and helping each other whenever we could. And I was infected: Dragonflies in slow motion was the "disease".

That year there was a big international dragonfly congress in Paris. Despite the high density of recognised experts there, we, Gunnar, Andreas and I, decided to go there to show the new recordings. Said and done. We packed everything and set off. The congress was to start in two days. It was just the time when Jonathan Waage had discovered that male Demoiselles remove the sperm in the female that a predecessor male had already put into the female and replace it with their own. A sensation for evolutionary research. And Ola Fincke reported of the Giant Damselflies from Panama, which create mini ecosystems in tree holes and float through the air as aesthetic magical creatures. "We would like to show some films..." was my address to the elegantly dressed director of the conference. The reaction was devastating. For half a year, it had been impossible to register a lecture – impossible. How naive we were. And when we tried to console ourselves in the hotel in the evening and then cockroaches scurried across the floor of our room, our mood was at its lowest. Even the gallows humour fell silent. The next morning in the panelled lecture hall of the museum, however, the surprise: "Monsieur! You, yes, yes, Monsieur, is it possible – your presentation? The scheduled speaker is suddenly fell ill. "When?" "In an hour!" A speaker had fallen ill and now a slot was free... and quite soon. Our presentation was supposed to be in an hour. Gunnar, Andreas and I into the car; Gunnar, Andreas and I into the hotel; Gunnar, Andreas and I back to the museum. Film projector set up – everything okay. No, we had forgotten the take-up reel. The lights went out. I began, for the first time in the history of dragonfly studies, to show slow-motion footage of the flight and behaviour of dragonflies. Andreas manually wound the film. The miracle happened. The silverbacks of the dragonfly scene were moved to silence. Everyone had never seen anything like this before. Some said: "These films...!" In slow motion, everyone could see how dragonflies move their wings, how they fly but also communicate something, signal. From then on, we received invitations from all over the world to film the very different dragonflies in slow motion – from Panama, Japan, France, Germany or the USA. And we went everywhere.

AMONG THE GIANT DAMSELFLIES IN PANAMA

The inhabitants called their island Barro Colorado Island in the dammed Gatun Lake of the Panama Canal "Dumping Island". Many animal species from butterflies to bats had already been studied here, ecological models calculated, and forest growth predicted. Ola Fincke had made the fantastic system of Giant Damselflies, which breed in tree holes, her research field. And that's what we wanted to film. Since it was very hot on the first evening, Gunnar and I hiked to a clearing on the shore of Gatun Lake, undressed and got into the water. Nobody had told us about the crocodiles – nor about the ticks that were everywhere on the shore in the grass. We had to pick off and pluck off over two hundred of them in the evening. To get rid of all of them, even the smallest ones, we occasionally went to Taboga, a holiday island in the Pacific. Ticks don't like salt water and leave the skin. In the evening, many animals appeared at the station: tapirs and beautifully decorated night butterflies, which were attracted by a researcher with light, rum and jam. A Chagas predatory bug was dangerous. Despite the greatest caution, it came into our room by crawling through a gap under the door. We were highly alarmed, as we had been warned about it, because it can be fatal to humans. But Gunnar took care of it. Now began a time of intense search for the Giant Damselflies. We combed the jungle for fallen trees. The larvae of the Giant Damselflies live in water-filled holes there, and the magnificent flying animals fly in front of them. "Georg, come, here is a goldmine!" echoed one day from the jungle. Ola had discovered a lying primeval tree, at whose tree hole two males of the Giant Damselflies were fighting. I immediately rushed to her call. This was indeed the place of places. For the next six weeks I was there every day and was able to film a little of their fascinating behaviour in slow motion every day: fights, matings, egg laying into the tree holes, flying in the wheel – every day a little more and sometimes nothing. The largest males, who often won this tree trunk with its water hole, had wingspans almost twenty centimetres wide. But we also discovered much smaller males, who sneaked up to females when the large territory owner was busy. They could also come to a pairing. The Giant Damselflies flying on the spot in patches of light under gigantic primeval trees were among the most beautiful things we have ever seen. Gunnar also examined many other dragonflies. Unfortunately, he had to return to Germany, as a replacement came Wilfried (*name changed*), a photographing friend of one of my exam candidates – he was very helpful, smart and could take great photos. But he also brought surprises into our jungle life again and again! After I had picked him up from the airport in Panama City after the time-consuming, usual security checks, we were so hot that we urgently looked for a refreshment. "Officers Casino" announced a sign on a long street at a bar for American soldiers. In immediately. We were let in friendly – but then chaos broke out. Right after entering, Wilfried swung his large backpack from his back, in an arc against a wall, on which hung a huge picture of J.F. Kennedy, and tore it down. It shattered into a thousand pieces. Disbelieving astonishment from the surrounding officers, a moment of absolute silence, then a roar that drowned out our: "Sorry, sorry!" Then a hasty retreat with a flight-like character, then again, the hot street and for the first time: "What was that?"

Over the phone, I had already told Wilfried about the ticks. Therefore, he brought a white overall for safety, good for spotting the little beasts – but the overall had wide slits at the pockets for the hands to pass through, through which the ticks also crawled. In droves. How many hours did I spend picking ticks out of his red hair and beard? Thus, saltwater bathing on the island of Taboga was particularly important for him. But it happened again on the way there. As soon as we landed on the island, we headed for the beach. It was very hot, and shade was necessary. Under a palm tree sat a young woman in a bikini with a four-year-old son. We headed straight for this tree shade. Wilfried, who had just drunk a can of cola, held the empty can and sandwich waste in his hand. Right next to the lady was a plastic basket. Unperturbed, Wilfried went to the basket and dumped his leftovers into it. "Oh, my god!" came the shrill voice from the woman: "Do you know what you are doing?" Wilfried had filled this probably well-off lady's bathing basket with his trash. With many apologies, we moved away, looked for another shady spot and I doubted again. When we first set up in the research station, I introduced Wilfried to the rules. Sleep here, mosquito nets definitely there, power charging there, kitchen there. "Oh yes, and you can help yourself – everything is in the fridge. You just have to clean up and wash up!" And that worked out pretty well, until one morning the station manager's voice rang out: "George, that's too much!" It sounded pretty shrill. What was going on? Nelly (*name changed*), the otherwise very friendly, American boss of the station was completely outraged. She thought we had to leave!

That couldn't be! I was in the process of making a film documentary about the largest damselfly in the world – also for television. The high travel costs. Our hope to uncover secrets and now this. The mystery cleared up quickly. Nelly had a birthday and had prepared her parents' loving birthday package for breakfast on the dining table – with flowers and cake and small gifts. Wilfried had sat down and eaten everything. What a disaster! The mood had been so good before. Only with Ola's intense advocacy, only with a lecture from me about bird flight, without aids, but with homemade cookies, were we able to convince Nelly that we could stay. But I stayed three days too long. My visa had expired, and I had to go to the offices in the tropically hot city of Panama City – sometimes in unlicensed taxis chased by other taxis – and I suffered from diarrhoea – it was hell. Whenever I stood somewhere for hours, I had to break off at some point. Finally, at the end of the day, I must have made such a suffering face in a long line that the lady at the counter asked me: "And you?" When I told her my story of suffering, she gave me the stamp for the visa extension without further questions. Soon after, I was on a plane in Panama City. But it barely got off the ground. I thought it was overloaded or the pilot was drunk. In Miami, however, I answered the concerned question of the Lufthansa employee, where I came from: "From the jungle!" I must have looked pretty bad and slept until Frankfurt without interruption – exhausted but happy, because I had the films with Giant Damselflies in my luggage!

Slum Battle in India

The meetings of the International Dragonfly Society were held at many exotic locations around the world, so that the participants on the excursions were able to see as many different dragonfly species as possible in their habitats. This time, in 1988, India was the destination, specifically the exotic temple city of Madurai in the south. It was not so easy to get there. When we, Gunnar, my son Olav and I arrived in New Delhi in the evening, we found out that our connecting flight to Madras had to be cancelled. So, we slept on the floor in the departure hall of the airport, along with dozens of other stranded people. But the next day it worked out: we arrived in Madras late in the afternoon. Luckily, we had an India expert, Mike, in our small travel group, who recommended a suitable hotel near the airport. After we had reunited our luggage, scattered by overzealous men, in a taxi, we reached the recommended "resort" totally exhausted. But this night was not restful either, because our room turned out to be quite open: behind a hand-painted picture, there was a large hole in the wall leading outside. Not only the humidity came in, but also insects. Mosquitoes, ants and cockroaches started their activities at dusk. We wrapped the bedposts with adhesive tapes in such a way that adhesive surfaces also showed to the outside and thus stopped the pests from climbing up the posts to us. Olav first crawled into a sea bag and closed the opening with gauze. But it became too hot and too oxygen-poor, so he soon started making alarming breathing noises. So out of the sea bag again and just put his head in a gauze hood. But that was also too hot, so he, like the other two of us, hardly slept because of the insect defence. We had a day of travel interruption in Madras due to a missing connecting flight. We wanted to use that. So, we spontaneously marched out of the hotel – slowly, because Gunnar had a knee injury. To the right or to the left? To the right! The path led straight into a very poor slum. We were stared at by ragged people and felt extremely uncomfortable. When a group of young men approached us – without moving a single inch to the side, our composure was finally gone. "Gunnar, can you walk quickly?" "Badly!" "Then let's go to the light over there quickly." And we walked towards the edge of the slum, which was bordered by a railway line. We managed to get there unhindered, finally climbed over the tracks and reached a road via a fallow field. And here, after a few minutes, a motor rickshaw came rattling along – and it stopped at our signal. "To the sea – we want to swim!" The driver understood that well, just like the acrobatic steering through the urban jungle of Madras, which soon closed around us. It's best to shut your eyes, so surprising were the curves, braking and evasive manoeuvres through the swirling chaos of other road users, holy cows and stalls, with which the streets of Madras were filled. Our driver suddenly stopped and said: "There is the sea!" But we saw no sea, only tens of thousands of black heads, bustling in one direction. We nodded hesitantly to each other and let ourselves be carried along by the crowd. "They surely don't all want to bathe." Olav noted when we had already walked a few hundred meters. And then: "I read in the travel guide: Be careful at religious festivals, they can easily turn into xenophobia. Is this something like that?" The situation slowly became tense. I looked at Gunnar. But he, as a great ornithologist, was looking through binoculars at vultures circling above us. I hesitated when I noticed that the attention of individual teenagers had turned to us. They stared at us; small stones flew at us. We looked at each other in dismay. "Inconspicuous retreat." I muttered. Then we turned around. But a group of young, black-haired men followed us. We accelerated – so did they. After about a hundred meters, several people in the crowd in front of us screamed: two men had gotten into a fight and were hitting each other with bamboo sticks. A circle had formed around it. Everyone was very excited. Old women clicked their tongues. But we squeezed past the tumult on the side and soon reached the street unhindered, where several rickshaws were standing. "To a restaurant, please!" Oh, how heavenly it was to eat Indian food under a huge fan... Even when we sat outside our hotel in the evening, drinking a cool drink on a bench, we found life wonderful again – despite insects in the night. The flight the next morning to the temple and congress city of Madurai was on time and we drove into the city with a shuttle bus after landing. Many beggars on the roadside, a sad sight. An American colleague apparently found this too much, as he wanted to fly home the very next day. Gunnar asked what the hotel was like. I said, like yesterday. He groaned. But when we opened the door to the air-conditioned room of a 4-star hotel, he fell onto the bed and said: "We can survive here!" And that's what we did.

Alone, we explored the huge temple in Madurai. It was magnificent to experience the intricate shrines and sacred figures, as well as the devout people, in the cool silence. The conference also lived up to its promise. Peter Miller, the English colleague who was on a one-year sabbatical in Madurai, had also ensured this. Discussing his research findings and especially many aspects of dragonfly life with him was very rewarding. We particularly discussed motivation as a behavioural drive. At that time, we could not yet know that behaviour is also switched on and controlled by hormone-driven gene activation. So, we mainly exchanged assumptions based on observations. We got additional time to discuss when our flight connection from Madurai to Madras was delayed by eight hours. Gunnar, Olav, and I wanted to go to Kathmandu in Nepal, because there was supposed to be a primeval dragonfly (*Epiophlebia*) living there that we wanted to film. In Kolkata, our onward flight to Kathmandu was completely cancelled that day because a minister needed the plane. This time, twenty dollars in the hand of an official was enough to get a remote office as a passable night's quarters. We spent the day in the rich but slightly dusty national museum and strolling through the centre of Kolkata full of colonial buildings and bridges overflowing with people. In Kathmandu in Nepal, we searched for *Epiophlebia* living there on a small river but could not find it. Instead, there were magnificent views of the Himalayas. However, before we started our return flight, we drove to a low-lying wildlife sanctuary. Due to a landslide, we had to stop for several hours, but then in the national park we were rewarded with rhinos and a huge tiger that crossed our path. On the return flight, the pilot made a detour near the roof of the world for our sake "The flight monitoring here is not so strict". When many passengers in the plane rushed to the Himalaya side, the plane tilted a little and the pilot said audibly: "Oops".

FUNNY DRAGONFLY RESEARCHER

That summer, I was able to film Black-tailed Skimmers, for many days at a large, water-filled sand pit in the north of Braunschweig. The males also hunt for females. These often come to the water to lay eggs when the sky is overcast or at dusk, because then they are not harassed as often by the males, who mainly fly in sunshine. If a male had caught a female, it often flew one or more loops with it. In the slow-motion shots, it became clear that each time a bright liquid squirted out of the female. We wondered what that could be. Eggs already fertilized by another male, water, or even sperm from a predecessor? That was indeed an interesting question. So, I wanted to catch these little drops in the air. I built a one a metre-wide, round disc made of black plastic film, which was attached to a bamboo stick. Then I followed the Black-tailed Skimmer males with my eyes. Every time they had grabbed a female, I sprinted into the shallow water and thrust my strange plastic disc into the air where the action was taking place. It was a fruitful morning, as about 8–10 pairings came about. Tirelessly, I ran into the pond and thrust into the air. It must have looked very funny. Out of the corner of my eye, I saw a rider who had stopped at the shore. "What are you doing there?" he asked. "I'm trying to catch the sperm of dragonfly males." His facial features slipped, he shook his head, turned his horse and galloped away. Obviously, he thought I was crazy. Yes, dragonfly research is often incomprehensible at first glance, but often aims at very important questions of evolution and ecology.

DAGMAR JOINED ME IN THE DRAGONFLY WONDERLAND OF JAPAN

I desperately needed help with filming the primeval dragonfly (*Epiophlebia*), this time in Japan. No one had time. Only Gunnar, but only for two weeks. In response to my many questions, a young woman – Dagmar – responded at a nature film seminar. She had time and also a pen pal in Japan. She could replace Gunnar. That was a very good suggestion, because from then on, we researched, lived and filmed together. Her flight to Japan was uneventful, if you exclude the adventurous stopover in Hong Kong. Very close to the skyscrapers, felt only fifty meters, the wings scraped past the glass towers of the old airport. The air when getting off was like hot cotton. After ignoring the towing attempts of an Asian man, she arrived in Japan – tired, but satisfied. So far, we had not been very successful in our attempt to film the ancient dragonflies. It was not easy, because they flew fast and in their search for food mainly in the tops of conifers. Even with telephoto lenses, we couldn't reach there. One of the helpful Japanese, Fujimoto, helped. He drove a lift truck extendable to six meters height right into the forest, right under the dragonfly trees! We were overwhelmed! The films were successful up there. Here too, they revealed the original and extremely economical flight style: unlike the modern dragonflies, the ancient dragonfly always flies with opposing movements of the two pairs of wings. This results in constant propulsion and the dragonfly is not slowed down with each wing beat and has to accelerate anew. It also doesn't have much power to move forward. Quick manoeuvres – no way. The landing was more of a crash. Only here on the cool mountain slopes could the poorly flying ancient dragonflies survive without competition with the modern, large, sun-hungry dragonflies. There was one catch, however, when filming up in the trees: Fujimoto didn't know how to get us down again. But after extensive reading of the instructions, he finally managed that too. Once he lent us his off-road vehicle for two weeks on the condition of also being allowed to pay for the gas! A few weeks later, Fujimoto was bitten in his garden by a poisonous snake, of which there are many in Japan. But he survived. Dagmar, by the way, had fallen asleep among green and flowering plants by the dragonfly stream right next to snakes after her arrival and had been luckier. She was not bitten. However, a few days later she had a decent scare: When we had later borrowed a small Honda car, she drove off briskly in the middle of Osaka when I shouted "Brake!" Suddenly a huge truck loomed in front of us. Dagmar had not thought that there is left-hand traffic in Japan. Fortunately, the driver towering over us had not overlooked us and had also braked. We drove to the ferry to visit the Dragonfly Park in Nakamura. The car ride was nice, but full of surprises, because we only had a Japanese road map. So, we stopped after the characters. We named the lookout with images like "teapot" or "little man with a walking stick" to find them together in the sign chaos of the streets. We moved slowly, but somehow, we got to the ferry and arrived at the Dragonfly Park in Nakamura the next day. There we wanted to film dragonflies with coloured wings, because we thought they would fly differently than dragonflies with transparent wings, which was later confirmed.

When Dagmar wanted to photograph them, she moved a little bit backwards and stumbled over a small wooden boundary of a ditch and fell backwards into it. She disappeared up to her upper body in the black mud. Startled, I immediately tried to pull her out. But then I had another idea: "Stay! Please!" and then I first took a photo of her. This aroused such incomprehension among Japanese visitors to the park that they scolded loudly, but at the same time also smiled. Japanese women are often elegantly dressed in public, usually in light colours and very friendly. But here something so unusual had happened that they had not often experienced, and they wondered how it would continue. Luckily, I had a very large rain poncho with me. Dagmar wrapped herself in it. I carried her to the car, and we drove into the mountains. Japanese mountains are full of clean streams and rivers and sometimes very lonely. Dagmar bathed and we washed everything. It was very warm and after two hours everything was clean and dry again. Luckily, we had bought a luxury pack of sushi and plenty to drink and so it became a really nice bathing and picnic afternoon – and we even filmed! On the day of departure, we wanted to be sure that the battery of the slow-motion camera, which had smoked on the outward flight, would not cause problems again. So, we stopped in a parking lot, took out the battery, put it on the ground and started to discharge it. Sparks were created – and a suddenly howling engine noise. A truck driver had observed us and had quickly started his huge truck. Maybe he thought we were handling a bomb – and when I look at the battery today, I have to admit that it also looks quite military, so olive green and long and round. This time, however, the battery only made us laugh.

A year later we were back in Japan. This time one of the big dragonfly congresses took place again. It was one of the best organized conferences with large, public lectures. Almost 1000 people came to one, especially children, to whom we were allowed to show our slow-motion film. However, the jumping frogs with their catapult tongue outdid the dragonflies. Also, on this trip there was a dangerous incident. This time we were traveling with Frank. He is an expert on everything to do with dragonflies and knows his way around the world. He was diving in a river to catch larvae to identify them when an earthquake surprised us. He didn't notice it in the water, Dagmar and I did on land. But it was not a big one and we continued our exploration trip on a mountain. But then it happened: We had walked a forest path; Frank ahead and Dagmar and I behind. To the right, the landscape fell off quite steeply. About thirty meters below us a waterfall shone. As often in Japan, it was one of those romantic landscapes that you want to fly into. I actually did: I had put my expensive telephoto lens in front of me on the edge of the path and looked around: When I turned around, I accidentally hit it with my foot. It tipped over and began to tumble down the slope, which was covered with many boulders between the pine trees. I immediately jumped after it and tried to grab the lens. I slipped on the rubble, fell and shot downwards faster and faster. For sure two to three seconds, I only saw blurred stripes rushing past on the right and left, until I hit a tree trunk – then a short rumble above me and the heavy telephoto lens crashed into my back. Good! Not good at all! Horrified, I looked down. The waterfall below me was not far away and I had fortunately come to a stop through a fir tree, noble fir! Carefully, I checked if I was injured, but didn't notice anything yet. I probably had a slight shock. Slowly, I crawled piece by piece upwards. Upon reaching Dagmar, who was very excited, I sank trembling to the ground. The tree trunk had abruptly stopped my right side of the body and it started to hurt.

221

Frank and Dagmar dragged me to the car and the ride to the hotel was no pleasure trip. In bed, I began to process my spontaneous, thoughtless behaviour. It had been very inconsiderate and selfish to risk so much for a glass-filled metal casing, no matter how expensive. I didn't want a Japanese doctor, having heard, mistakenly, about all sorts of traditional medicine. I still feel the consequences of this fall today. When the pain started and a typhoon made the high-rise sway back and forth with us on the twentieth floor, romantic Japan turned into a nightmare that night. After a few days, however, the pain became bearable, and we were able to travel to Hokkaido in the north of Japan. There we wanted to see the famous Japanese cranes or "Tancho" symbols of loyalty, happiness, love, and long life. In addition, Dagmar needed new species of dragonflies for a study of their activity distribution for her thesis.

In Hokkaido, Hidenori was researching dragonflies. We wanted to visit him. His ingenious experiments were exemplary. He had invited us to Kushiro. So, we also attended a scientific session of his working group. It was very interesting to be able to follow the winter ecology of insects in this cold part of Japan. How they produce antifreeze at great cold to prevent freezing and how dragonflies can be active in the snow was new to us. We were invited to dinner – in a fancy restaurant. First, there was a fish platter. Dagmar and I were served first and – almost ate it up because we were hungry. How great was our surprise when it was then passed on to the other participants. What a faux pas on our part! We had practically left only bones for the others. When we then ordered an expensive French wine, the embarrassment was perfect. Our hosts didn't show the slightest sign and we apologized the next day. Foreign customs and our insensitive behaviour did not match that evening.

At a nature film congress in Japan another year later, we were to show our dragonfly film. We were sitting in the middle of the audience. It was an impressive experience, as we had our seats in the middle of a group of monks. These half-naked, reverent men had never seen such slow-motion shots of flying dragonflies and often expressed their astonishment loudly. Many vowels, many ahs and ohs – it was very impressive for us to experience the reactions of these holy men up close. There must have been something spiritual going on in them when they could see and understand the otherwise invisible movements of the beautiful insects so well. When we were able to answer questions on stage afterwards, we were very happy. Our joy increased when Kiyoshi suddenly approached from the back door, came up to us on stage and sang "Akatombo", the famous Japanese children's song about a red dragonfly, with us and the audience. He had driven over 400 km, spent two days to see us!

As intoxicating as this stay in Japan was, the last return flight was more like a nightmare. On the last day of our trip, we had been invited to a buffet by a large Japanese company that manufactured media devices. When we entered the dining room, our eyes nearly fell out of our heads at the artistic presentation of the food. A Japanese businessman indicated to me, next to a tree where many sushi shone, that I should help myself. I did so out of politeness, but also out of hunger. But somehow, the sushi tasted a bit soapy. I didn't think much of it and continued eating, accompanied by the pleasing smile of my neighbour. But I hadn't reckoned with the consequences of this meal: at three o'clock in the morning, an earthquake began in my stomach and intestines. When the bus to the airport arrived the next morning at seven o'clock, I could barely drag myself in. On the way in the bus, there was a person check, which I spent in the fortunately available toilet of the bus. At the airport, I groaned to Dagmar: "I can't go with you." She then rushed to a pharmacy and returned after a quarter of an hour with a medication. Of about twenty Japanese and fortunately also partially English labelled remedies that had been presented to her; she had guessed this one. And it worked – tolerably. At least I was able to pass the various counters of the security check and baggage handling soon after taking it. Then I sat in the crowded jumbo jet with a narrow seating arrangement measured according to Asian body sizes. When the stewardess came around, she discovered me and my condition. She asked a Japanese woman sitting next to me to move to another seat, so I could lie down, my legs on Dagmar's lap. After four hours, somewhere over Siberia, I felt so sick that the desire to jump off became overpowering. I had to go to the toilet constantly, which the surrounding passengers found at least strange. But finally, after eleven hours of flight, we arrived in London. There, a deep sleep overcame us. If a cleaning lady hadn't knocked over a metal bucket near us, our connecting flight to Hanover would certainly have left without us.

SWAMP AND STONE STEPPE IN SOUTHERN FRANCE

By chance, Gunnar from our dragonfly working group discovered a gem in southern France, a fast-flowing stream, the Canal de Vergière at the entrance to the Crau, a unique stone steppe in Europe southeast of Arles. In this clear, herb-rich water, we counted over 40 species of dragonflies. From then on, this area was a frequent target for our research projects. In addition to the richness of species and the usually beautiful weather there, another advantage was that the stream was very shallow, so that one could stand in the stream or sit on a tripod researching the dragonflies. Many students also came along on excursions here. The nature of the Crau and the nearby Camargue thrilled everyone – the intense sun, which had also given Van Gogh a hard time and sometimes too much wine brought certain complications with it – and love… One candidate fell in love with a shepherd. It was funny when in the treeless Crau as the only elevations from one side a flock of sheep and a shepherd with a wide-brimmed hat and from the other side a lonely person approached each other. As soon as the two silhouettes had become one, they disappeared for a while for a classic shepherd's hour… Another student fell in love with the landscape and the animals and stayed there forever. In the middle of the Camargue, she moved into a small house and researched this huge swamp ecosystem at the Tour du Valat research station. We fought the great midday heat by lying in the canal or, when it got too bad in August, driving to the nearby mountains, the Alpilles, with shady forest.

Another problem was car break-ins. During a soccer world cup, we were watching on TV when our landlady came to us excitedly: "Monsieur, Monsieur – votre voiture!" We ran out to the street and saw the mess: the car doors were open and most of the car's inventory was scattered on the street. A lens was missing, a pair of new sneakers as well, and a few other small things. Oh, that was annoying. We also experienced such car break-ins in Arles and in the Alpilles, at St-Rémy-de-Provence. We were told that the culprits were probably teenagers from Paris who were financing their vacation this way. But what was that compared to all the beauties of the landscapes, the romantic cities with the colourful, lively markets. Sitting there, drinking a coffee and listening to chanson singers was grandiose. Southern French flair was also present at many other festivals. Once, I caused chaos at a sheep drive through St. Rem de la Provence. In order to better photograph the thousands of sheep, I climbed onto a garbage container. This container had a large lid that swung open to the side, on which three small children were already standing. When I also stepped on it, the lid flipped down with us. We fell onto the pavement... The children screamed, adults pulled them up with horrified faces and frightened exclamations: "Monsieur, qu'avez-vous fait?" – "What have you done?" was the mildest. When I was finally lying down there alone, a young woman bent over to me and asked me how I was doing.

A music festival in Arles particularly appealed to us. Everywhere in different parts of the city, bands were marching through the streets. One group was particularly original: the musicians were dressed as insects. We, as insect researchers, of course, liked this very much. We followed them and were also thrilled by their music. When they settled down in the centre on the steps of the cathedral and played music there with great verve, an obviously high-ranking clergyman came and sent them away, accompanied by boos. They continued to play in front of a museum.

For the Second German Television, we made a television film about the behaviour of the dragonflies here at a stream in the Crau, and the BBC also came with a film crew to Southern France to film dragonflies with us. The highly paid camera people, however, were not satisfied with simple wooden bungalows on our campsite. They were okay for the equipment, they said, but they themselves wanted to stay in a 4-star hotel for the night. One must understand this, as these teams are used to living in expensive lodges and hotels worldwide, because, for example, in African countries there are hardly any other possibilities for restful overnight stays. So, they have high budgets for this and why should they give any of it away. However, they also worked highly concentrated and very effectively from morning to evening. We learned a lot from them. As is usually the case with nature lovers or film colleagues, this cooperation was also extremely friendly. We could only dream of the production costs that the BBC used here. Because the sunflowers were not in full bloom at the time, it was decided to fly an extra film crew from England to southern France for a later shot, just to catch up on this one shot.

METAL-WINGED IN THAILAND

The most beautiful dragonflies in the world? I had already filmed them – fifteen years ago in Panama. Yes, but these in Thailand would have metallic shiny hind wings, which shine in very different colours depending on how the light falls on them. Dagmar, Jan and I are easily convinced when it comes to making a trip. And Thailand with an ancient jungle … Okay, when is the season there? And half a year later we sat with André in a small office of a car rental company in Phuket. André had been here in Southeast Asia several times and was familiar with the mentality of the people and the peculiarities of nature here. We had brought him a slow-motion camera and explained it to him. So, we quickly learned from each other's experiences. The very next day we drove about eighty kilometres to the Kao Sok National Park. From the parking lot to the spot with the metal-winged in the river it was another three kilometres. Hiking there and carrying everything was not easy due to the humid heat – Jan, at eight years old, found it less difficult than me. Now began a breathtaking hunt for perhaps the most beautiful dragonflies in the world. Following their fast flight with the camera was a real challenge, but the three of us, Dagmar, André and I, managed to capture their behaviour in slow-motion shots. These revealed that the males, only they are so conspicuously coloured, keep their coloured hind wings still during flight and present them to rivals and females. They then fly only with their forewings. Who displays the most beautiful hind wings and the best flight, gets a territory and mates. Either Dagmar or I would go with Jan, accompanied by the concert of the gibbons, on an excursion through the jungle or to the nearby waterfall to swim with local children, where the fair-skinned and blonde Jan was the exotic one. On the third day, Jan sat with a red face on the bank of the river and didn't want to do anything. He had a fever, quite a high fever. So, I ran back to the gate. Sweating and panting, I arrived there, headed for a group of men who obviously belonged to the national park. I described the situation and handed them twenty dollars. With three mopeds, they set off and immediately went to fetch our three. Jan was able to hold on well even during the bumpy ride and the big André sometimes had to get off uphill – but eventually all three arrived back safely. In the hotel, a tourist doctor came in the evening with a magic remedy, because after another day Jan was healthy again and could already go on a diving excursion to an offshore island the day after next.

ON THE OKER

The most beautiful experiences with dragonflies, however, we had on the Oker in Northern Germany. The Oker is a small river that comes from the Harz Mountains and is about 120 kilometres long. North of Braunschweig, it is about 15 meters wide and not much deeper than a meter. At particularly exposed spots, it can also be 2 meters. In largely natural meanders, the river winds its way through a diverse floodplain at a flow rate of 0.5 to 1 meter per second. The bank is lined with willows and other softwoods as well as bank plants such as water meadows, Reed Mannagrass (*Glyceria maxima*) and in the water grow Arrowhead (*Sagittaria latifolia*), Yellow Pond Lily (*Nuphar lutea*) or Flowering Rush (*Butomus umbellatus*). Underwater, the many leaves of the Watermilfoil (*Myriophyllum*) shimmer. From the small village in the north of Braunschweig named Volkse, we paddled upstream by canoe and looked for good spots where many water plants promised a lot of dragonfly activity. At such places, we anchored the boat on the bank, prepared the cameras and let the day unfold. The process was always similar. In the morning, the Demoiselles gathered in the bank plants and bushes and began to hunt. This was the first challenge for them and us: following the rapid damselflies with the camera and filming them catching small insects was only possible with quick reactions and generous running of the camera. Gradually, the males slid down to the water and began to fight for good territories. Everywhere their blue wings flashed and were swung like signal flags towards the opponents. Up and down, back and forth led their flight paths and now and then there were also arrow-fast and curvy chases. This lasted a few hours until around noon the females, hidden by their camouflage colouring, detached themselves from the bank plants and flew to the water. Then the blue damselfly males outdid each other. With an increased beat frequency, they approached the females in a courtship flight. It was very exciting whether and when a female accepted the suitor. But everything was very orderly and peaceful. That was about to change. Dagmar's grandfather gave us an unexpected discovery – and this is how it happened: It was his birthday and Dagmar said: "Let's stay longer and then visit him afterwards. Then we don't have to drive back and forth". Said and done. We stayed longer. This turned out to be a stroke of luck. Around 7:00 PM it slowly became twilight; it was already September. The Demoiselles were still active, because it was not cold. But what we saw now was completely unexpected: A behaviour of Demoiselles that we had never observed before developed. The males no longer courted, no, they now pursued the females at the sleeping place, where they all came together. They pounced on them wherever they could and began to mate everywhere, even though it was slowly getting dark. It was pure chaos, and we were completely stunned. Our peaceful Demoiselles had turned into aggressive, archaic transformed into seeming robbers. It was incredible how our orderly, painstakingly constructed image of the peaceful dragonflies changed: this was the alternative reproduction process that we had just discovered there. Thanks to the birthday of Dagmar's grandfather. At the evening celebration, we certainly did not always leave a concentrated impression.

In the Air Almost Like a Dragonfly

New perspectives are the be-all and end-all in modern nature film. So, we got ourselves a drone to be able to depict the habitats of the dragonflies from the air as well. Luckily, we had a capable drone pilot in our son Jan. He didn't mind making time-lapse recordings of the approaching morning in the foggy meadow at five in the morning, even in cool temperatures. Yes, he even enjoys flying in difficult terrain between trees or under bridges to get unusual views. This was a great gain for our filming. But when he no longer had time for study reasons, we decided to fly the drone ourselves. That went wrong several times. One difficulty in operating the drone is that you have to keep an eye on three battery charges: the battery level of the mobile phone with which you control, the battery level of the drive directly on the drone and that of the controller. If the charge on the two drone batteries becomes too low, the drone automatically switches to return function, which means it rises sixty meters high and then immediately flies back to the launch point. That day I had let the drone fly along the bank of a river bend of the Oker and had not paid attention to all three batteries. Then I heard a high buzzing coming closer. Oh, my goodness, the drone came rushing in, stopped above me and then lowered until it got stuck about six meters above me in a tree. The rotors were braked by branches and stopped. There it hung dangerously over the surface of the river. What to do? Just leave

228

the expensive piece hanging and let it fall into the water? No, that was out of the question. So, I rushed to the nearest hardware store ten kilometres away, bought a saw, a fishing net and a step ladder. After half an hour I came back. Oh, my goodness – the drone was gone. My gaze slid down the trunk of the tree and stopped at the edge of the steep bank. There – I couldn't believe it, there it was, the drone, right at the end of the steep edge above the Oker. A gust of wind would be enough to let it fall into the water. I approached carefully and managed to grab it. What a stroke of luck. When I told Jan the story on the phone, he just said laconically: "Why so?" But even this experience did not stop me from forgetting the battery levels two more times and repeating such disasters. Once the drone got tangled up in an oak tree above me. As I stared up for minutes trying to think of a rescue, it suddenly fell down and landed next to me in a blackberry bush – intact. It's not easy to move through the air like a dragonfly.